Technology

Made Simple

The Made Simple series
has been created
especially for self-education
but can equally well
be used as
an aid to group study.
However complex the subject,
the reader is taken
step by step,
clearly and methodically,
through the course. Each volume
has been prepared by experts,
taking account of
modern educational requirements,
to ensure the most
effective way of
acquiring knowledge.

In the same series

Accounting
Acting and Stagecraft
Additional Mathematics
Administration in Business
Advertising
Anthropology
Applied Economics
Applied Mathematics
Applied Mechanics
Art Appreciation
Art of Speaking
Art of Writing
Biology
Book-keeping
Britain and the European
  Community
British Constitution
Business and Administrative
  Organisation
Business Economics
Business Statistics and Accounting
Business Calculations
Business Law
Calculus
Chemistry
Childcare
Child Development
Commerce
Company Law
Computer Programming
Computers and Microprocessors
Cookery
Cost and Management Accounting
Data Processing
Economic History
Economic and Social Geography
Economics
Effective Communication
Electricity
Electronic Computers
Electronics
English
English Literature
Financial Management
French
Geology
German

Housing, Tenancy and Planning
  Law
Human Anatomy
Human Biology
Italian
Journalism
Latin
Law
Management
Marketing
Mathematics
Modern Biology
Modern Electronics
Modern European History
Modern Mathematics
Modern World Affairs
Money and Banking
Music
New Mathematics
Office Administration
Office Practice
Organic Chemistry
Personnel Management
Philosophy
Photography
Physical Geography
Physics
Practical Typewriting
Psychiatry
Psychology
Public Relations
Public Sector Economics
Rapid Reading
Religious Studies
Russian
Salesmanship
Secretarial Practice
Social Services
Sociology
Spanish
Statistics
Technology
Teeline Shorthand
Twentieth-Century British
  History
Typing
Woodwork

Technology

Made Simple

Professor Don McCloy

Made Simple Books
HEINEMANN : London

Made and Printed in Great Britain
by Richard Clay (The Chaucer Press) Ltd, Bungay, Suffolk
for the publishers William Heinemann Ltd,
10 Upper Grosvenor Street, London W1X 9PA

British Library Cataloguing in Publication Data

McCloy, Don
  Technology made simple.—(Made simple
  books, ISSN 0265-0541)
  1. Technology
  I. Title   II. Series
  600   T45

ISBN 0-434-98596-1

Editorial: Robert Postema, F. G. Thomas
Production: Martin Corteel
Text diagrams: Reg MacClure, Reproduction Drawings Ltd
Cover illustration: Derek Hazeldine Associates

Foreword

The explosive growth of technology requires today's education to provide people with a flexible outlook—one that will make them adaptable and prepared for change. The school curriculum is now under close scrutiny to see what can be done. In higher education, delayed choice of specialisation is becoming more and more common in degree courses; and continuing education is a growing area within which adults, both lay and professional, are being updated or redirected in their careers.

There are two good reasons why Technology should be taught in the schools. First, to prepare young people to live in a technological society; secondly, to prepare some of them to work as technologists. These were highlighted in the Finniston Report on the engineering profession in the UK. It argued for a strengthened technology base *and* for a technologically aware public. A successful country nowadays needs not only top-class practitioners of technology, it also needs a public with an awareness of the importance of technology and an ability to make intelligent comment on technological issues. Our educational system must respond at all levels to provide some means of inculcating a greater technological literacy in society.

National and economic needs aside, however, there is a lot to be said for the educational process inherent within the subject of Technology—a process that does not push knowledge into watertight compartments such as physics, mathematics, chemistry, English and so on. Technology draws together these various disciplines. It is a problem-solving exercise, and today's technologists have to apply their knowledge of mathematics, of science, of economics, of manufacturing processes, in this exercise. Technology integrates these disciplines in its problem-solving activity.

This need to draw on knowledge of many different kinds is reflected in the definition of Technology: the systematic application of scientific or other organised knowledge to practical tasks. This definition also makes it clear that Technology is concerned with doing things and providing solutions to problems. Is there not nowadays an overemphasis on the acquisition and regurgitation of facts? Education has been accused of being not far removed from that advocated by the character in Dickens' *Hard Times* who said: 'Now what I want is facts. Teach these boys and girls nothing but facts. Facts alone are wanted in life. Plant nothing else, and root out everything else.'

A well balanced education should certainly include analysis and the gathering of knowledge, but it should also include the development of the creative skills and the problem-solving skills. We need to pay more attention to synthesis. Our young people need to be trained to make judgements and intelligent decisions, often in the fact of inadequate knowledge. How often do we ask 'Look at this and tell me how it works', rather than 'I need this job done, tell me how to do it'? For most of our lives we have to deal with the latter situation. We have to solve problems using all the knowledge and skills we can muster. Isn't that what life is all about?

This book is an attempt to introduce the reader to the most basic knowledge that the technologist needs. But it also tries to show how this knowledge can be applied and how its application might impact on society. Although a basic knowledge of calculus would be helpful, it is not essential, and the intelligent reader with no mathematical knowledge should feel at home throughout the large majority of the pages.

The book is principally intended for self-study, for the educated layman who wishes to be more aware of technology. However, it will also be of value to schools, colleges and universities. It is particularly relevant to the ordinary and advanced Technology subjects offered by the GCE and CSE Boards. It is recommended as supportive reading for the many other technologically orientated subjects such as Computer Studies, Design, Applied Mechanics and Craft, Design and Technology. In addition, it will be useful to teachers of Mathematics, Chemistry, Physics and Biology who wish to enrich their subject by the inclusion of an element of technology. Teachers of Level 3 BTEC courses in Engineering should also find useful material in the book, particularly in its method of presentation. Finally, because of its level and breadth, the book could also be used for the technological studies content of degree courses in non-technological disciplines.

This was not an easy book to write, but the author's contacts with the schools through the Northern Ireland GCE Board and the Northern Ireland Science and Technology Regional Organisation (NISTRO) at Ulster Polytechnic convinced him that it should be done. The author would like to acknowledge the helpful advice received from Mr Roger McCune, Executive Officer of NISTRO, and from Dr Martin Brown, Head of Physics at Methodist College, Belfast.

The author is also indebted to his secretary, Mrs Estelle Goyer, for her sterling and patient endeavours. She and Reg MacClure, who did the excellent drawings, added a touch of sanity and restraint when the author's ideas overstepped the bounds of credibility.

D. McCloy

Contents

3 Technology Extends Man's Senses and Communication   71

10   Technology and Society

1
The History and Development of Technology

1.1 Introduction

Technology is as old as man. The word derives from the Greek words *techne*, meaning an art or skill, and *logia*, meaning a science or study. Thus the literal meaning of the word technology is the study of an art or skill. But this is a very general definition and does not emphasise the applied nature of technology. My favourite definition, attributed to Kenneth Galbraith, is that technology is the systematic application of scientific or other organised knowledge to practical tasks.

Man first became a technologist when he learned to take advantage of the materials and the natural phenomena of the physical world. When he discovered that a bone or a stick could be used to kill animals and to move rocks he became a tool-maker, and tools are the trademark of the technologist. This new-found technology was put to use in assisting Man to meet his basic needs of food, shelter and clothing. He would kill his prey with clubs, skin them and butcher them with flint knives. He was able to use the skins for clothes. He was able to fell trees and to move boulders to improve his shelter.

These embryonic technologies may have been crude but they signified a new outlook by Man. No longer would he be pushed about from pillar to post by nature. No longer would he live from moment to moment, reacting to each circumstance as it arose. The birth of technology changed Man to a problem-solving animal—to someone who deliberately brought about change. The invention of the tool, however simple, indicated planning, for when one makes a tool one has some use in mind. Problem-solving became a distinctive feature of Man and his tools helped him solve the problems. For example, a daily problem was how to get food. The solution was to use the flint tools, the sticks and other elementary hunting weapons that he had developed. This problem-solving capacity set Man aside from the lower animals who rely on instinct and learning. But it is not entirely unique to Man, for the higher animals, the apes, have demonstrated elementary problem-solving abilities, mainly by trial and error. For example, monkeys have shown that they can discover how to reach a banana by piling up a stack of boxes.

This developing ability to solve problems and to make tools accentuates another major difference between Man and the lower animals. Most animals have developed physical characteristics suited to a particular specialisation. The beak of a bird is a specialised tool for food gathering. The shark's shape is specialised towards fast movement in the water. The tiger's stripes are a specialised form of camouflage.

Man's specialisation is problem-solving or brain power and this has allowed him to outpace all the animals in their own naturally evolved specialisms. Indeed, technology is often referred to as the extension of Man's capabilities, and we shall discuss this aspect in more detail later. One obvious example should suffice to make the point. Our capability of sight has now been extended outwards through the telescope and inwards by the microscope. We can photograph terrain from reconnaissance aircraft at a height of 10 kilometres and pick out detail as small as a motorcar. We can see in the dark using infrared telescopes. Thus Man's specialism of problem-solving makes him ruler of the animal kingdom.

1.2 Methodology

A methodology of technology developed and was used, albeit unconsciously, by early Man (see Fig. 1.1). Basically one perceives a need: there is a problem. In order to solve this problem effectively one should prepare several solutions and then choose the best. But how do we know which is the best? There are often many conflicting aspects to be considered when making this decision.

Take an example that confronted early Man: how to capture prey? He might have considered two alternatives; he could run after it and club it, or he could make it fall into a pit. Which is the better solution? The first method is dangerous and exhausting. The second method could also be exhausting for a pit has to be dug; but this work could be shared by others and it would not be so dangerous—or would it? How, for example, would the animal be driven to the trap? Thus we can see a lot of pros and cons emerging once we give thought to these methods of solution. Which solution we consider the better would depend on one's attitudes to danger and to work, and this would vary from individual to individual. Thus in the decision-making stage we have to take into account all aspects of the problem and attempt to put the importance of each in order of priority. This is known as **'weighting'** and we shall hear more about it later.

The culmination of the technological method is the adoption and the construction of that method of solution which has been decreed to be best. But this is not the end of the story, for it is often found from operating experience that a chosen solution is not as good as had been predicted. For example it may be discovered that more people than animals are falling into the pit! In these cases it is 'back to the drawing board', and assumptions made during the earlier stages have to be re-examined.

There was a growing awareness of this technological method and with experience Man became more adventurous in his solutions to problems. The effectiveness of his solutions improved when his knowledge grew and he was able to consider more alternatives. For example, the discovery of fire gave him a new means of driving his prey into the pit we mentioned earlier. And again his study of the beasts led to the flash of inspiration

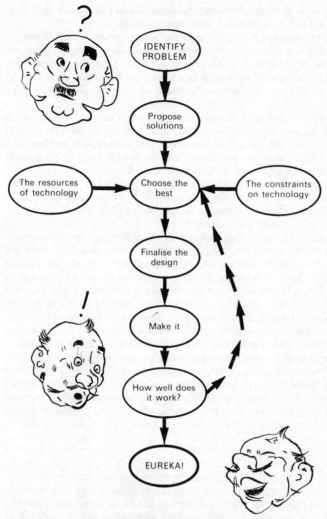

Fig. 1.1. The technological method.

that, rather than hunt them, a better solution to the food problem would be to tame them and breed them.

1.3   Empiricism and Science

Now as this knowledge of techniques and methods of solving problems grew, Man had a new problem—that of passing the information on. In the earliest days this was done by example, and the child would learn from the father. This is known as the **empirical approach**. In this way new generations did not have to start afresh to each problem. Rather they took their predecessor's solutions and refined them where possible.

In this way there was, in the true Darwinian sense, an evolution of solutions. Tools gradually improved by trial and error—the equivalent of the survival of the fittest of the flora and fauna. So technologies grew, each solution being a refinement of an earlier one; but occasionally a completely new solution arose from a newly discovered phenomenon or material.

This empirical approach continued right up to the Middle Ages when the rapidly expanding world of science came to the technologist's aid. **Science** is simply the quest for knowledge, and **the scientific method**, like the technological method, has several clearly defined steps. First several examples of some phenomenon are studied; then some hypothesis or explanation is proferred; next this explanation is tried out on other examples of a similar phenomenon; finally, if it seems to work reasonably well, that particular explanation is adopted, the particular phenomenon is understood and Man's fund of knowledge has increased.

For example, Newton's observations of falling objects, especially apples, led him to think deeply about the phenomenon of falling. His theory of gravitation was the hypothesis which he hoped would be the correct explanation for this phenomenon. He then tested this theory by considering the moon and planets as falling objects and was able to show that his theory of gravitation predicted their motion with reasonable accuracy. Thus Newton's Law of Gravitation was accepted as a scientific fact.

Now these new scientific facts increased the storehouse of knowledge that Man could invade when considering alternative solutions to his problems. He was able to apply science in his technology and a whole new field of **Applied Science** began to develop rapidly, especially in the middle of the nineteenth century. This impact of science on technology is best considered by examining the development of the various areas of technology, but we shall have to restrict ourselves to a consideration of the growth of Man's knowledge of energy and of materials.

1.4   Energy

The earliest men and women relied on the brute force of their own muscles. They did not realise that they were surrounded by many natural phenomena and resources that would provide them with energy greatly in excess of their own capability. Today we are familiar with these energy producers: the wind, the tides, running water, the heat of the sun, natural fuels, hot springs and so on. In fact, we live on a planet that is bathed in an unimaginable stream of energy emanating from our sun. The heat and light energy coming from the sun every second equals the power of a million million Hiroshima atomic bombs! About one part in two thousand million of this tremendous energy output reaches the Earth. Here some is reflected, some heats the atmosphere and causes our winds and our weather, and some gets through to warm the ground and make the plants grow. If we could only make use of all of this solar energy we could supply every human being with energy equivalent to each of us burning 100 tonnes of coal a day. Put another way, if all the world's fuels were gathered together and burned at a rate to match the sun's output, then they would only last four days.

But this ideal is a long way from being achieved, and it was extremely remote when modern Man emerged about 25 000 years ago. Then when power greater than an individual could provide was needed, it was necessary to arrange for the concerted effort of many people. This was greatly enhanced when Man learned to domesticate and harness draught animals like the ox and the horse. The completion of such huge tasks as the pyramids indicates just what is possible by human power alone.

Machines

Then there was the gradual discovery of devices to aid human muscle. Of mechanical devices for assisting Man's strength the **lever** is the most important. This transforms a small force acting through a long distance into a large force acting through a short distance. The origins of the lever are unknown, but we suspect that even primitive Man must have been aware of its use and may have used it for moving heavy objects. Man soon learned other tricks for getting the best from his limited power output. The engines of the human body can only do a certain amount of work in a certain time, and it was therefore necessary to slow the work rate involved in any task to this limit. Five basic machines evolved to help in this respect. In addition to the lever these were the **wedge**, the **screw**, the **pulley** and the **wheel and axle**. We shall hear more of them in Chapter 2.

It is interesting to note that with these machines, as with many other areas of technology, the earliest developments did not have the benefit of any scientific knowledge but were empirical and grew on a trial and error basis. The lever was not really understood until the great scientist Archimedes studied it some 250 years before Christ. It was he who made the somewhat ambitious statement: 'Give me but a place to stand and I can move the world.' All five basic machines were explained and related in the first century AD by another scientist (or natural philosopher as they were then known). He was Hero of Alexandria, the most voluminous writer on mechanics in the Alexandrian age.

Most of these machines have been used in ancient works. The wedge or slope was used extensively for raising great blocks of stone to a higher level. Although it may have been too exhausting to lift a block of stone through a single metre vertically, it was not too exacting to move it through the same vertical height up a gentle slope. The work of lifting the block vertically one metre can be done at a gentler rate by slowly moving it up the incline (see Chapter 2).

Thus the machines allowed Man, with his limited work rate ability, to do prodigious tasks. The lever was used in Greek beam presses as early as 1600 BC, to squeeze the juices of olives and grapes. Screws were used in oil presses in Pompeii. The wheel and axle that transformed a small force at the circumference to a large force near the centre were employed in early treadmills.

These machines allowed Man to adapt the task so that it could be accomplished at his relatively slow energy output rate. But there were occasions when it was necessary to produce a very high rate of work, and in order to achieve this it was necessary to develop devices that could release energy very quickly after having been charged with energy by manual effort. One such example involves gravitational energy. A man could slowly roll a large boulder to the top of a cliff and all the while it is accumulating gravitational energy or **potential energy**. By pushing it over the edge it can release this energy very quickly on an unfortunate animal standing underneath. Another example of a means of accumulating energy is the spring, and its first known use was in the bow for shooting arrows in hunting and in battle. Here the bowman could do the work of stretching the bow as slowly as he pleased. But when the string was freed, all this accumulated work would be released explosively in sending the arrow to its target. This knowledge was later to be applied to more and more complicated systems of warfare such as crossbows and catapults.

The most rapid changes in Man's use of energy did not, however, come until manpower was replaced, firstly by the power of draught animals and later by the power of water and wind.

Animal Power

Records from a period around 2000 BC. show that both dogs and horses were already being used to work for Man. The ass was used as a pack animal in upper Egypt at the beginning of the early dynastic period around 3000 BC. Donkey mills were used as early as the fifth century BC to crush ore in Roman silver mines and, slightly later in the third century BC, to grind corn in Greece. And, of course, the willing ox has for thousands of years assisted us in transport and in agriculture.

Besides being able to work tirelessly for long periods these animals could achieve power levels well in excess of Man's capabilities. For example, a horse in good harness can exert a pull up to 15 times greater than a man. Indeed, the horse's abilities were so greatly admired that its name was given to the unit of power, the horsepower.

The use of draught animals was only one small step in Man's exploitation of nature's resources. It whetted his appetite, but when he looked at the abundant evidence of the tremendous power of the rushing river, the waterfall and the hurricane, his ambitions grew.

Wind and Water

The sail was the first mechanism to harness natural power. It exploited the power of the winds and indeed was the first non-living **prime mover**. (A prime mover is a mechanism, or organism, that converts a natural form of energy into another form capable of providing motion.) The origin of the sail is obscure, but early drawings show square sails in use on Egyptian boats. Who knows what inspired the first boatman to use a sail? Maybe it was a leaf blown in the wind; maybe it was the wind pulling at his cloak?

So wind was Man's first ally in the quest for increased power. Water-power was to be the next, and in the hilly regions of the Near East the first waterwheels developed a few hundred years before the Christian era. These early wheels had a vertical axis and a horizontal wheel whose paddles caught the edge of the current of a fast flowing stream (Fig. 1.2). The stream turned the paddle-wheel, and its shaft was often connected to a millstone for grinding flour.

Mill wheel

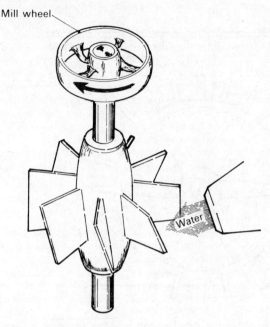

Horizontal waterwheel

Fig. 1.2. The Norse mill.

These mills were to become known as Norse mills because of their later widespread use in northern Europe. The Vitruvian watermill was of a different design, and was proposed by Vitruvius in the first century BC. There are still many of this type of mill to be seen in the countryside and their distinguishing feature is that the axle is horizontal and the wheel vertical. In the earliest Vitruvian mills the lower part of the wheel was submerged and was turned by the force of the current (Fig. 1.3). This was known as an undershot wheel. A wheel of about 2-metre diameter could develop about 2 kilowatts and could grind about 200 kilograms of corn every hour. (We shall define the kilowatt (1000 watts) later, but in the meantime, to give you a rough idea, a man shovelling coal onto a lorry generates, on average, around 50 watts.)

These mills made a great impact in many areas of industry. They were to be used for ore-crushing, for sawmills, for operating the bellows of furnaces and so on. They were very popular in England, and the Domesday Book tells us that there were 5624 water-mills in use south of the Trent in 1086.

The windmill is not nearly so ancient as the watermill, and it was not until around 500 AD that a Persian farmer had the bright idea of turning

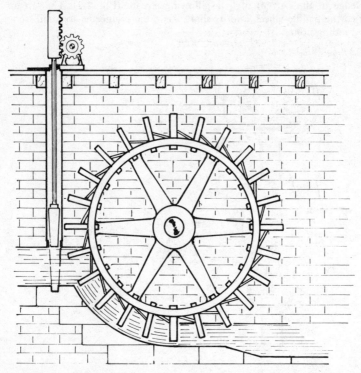

Fig. 1.3. An undershot Vitruvian watermill.

the Norse waterwheel on its head so that a current of air, rather than water, would make it revolve (Fig. 1.4(*a*)). In the Western world, on the other hand, windmills seem to have been inspired by the Vitruvian waterwheel with axis horizontal and wheel vertical—the sort of windmill that Don Quixote took a dislike to (see Fig. 1.4(*b*)).

The first windmills were used for grinding corn but their use grew to include water pumping, hoisting materials from mines and driving saws. By modern standards they were not powerful prime movers. Theoretical studies show that they could at best have generated only about 25 kilowatts. Indeed, it seems that the power of the average windmill was probably only about 3 to 6 kilowatts, which was roughly the same as the power of a watermill. Thus Man was still limited in power and he began to look for other natural sources of even greater potential.

The Steam Engine

This new source of energy was to come from the interaction of heat and the atmosphere, and it was at this stage that science began to make a considerable impact on the growth of technology.

These early machines relied on the principle that 'nature abhors a

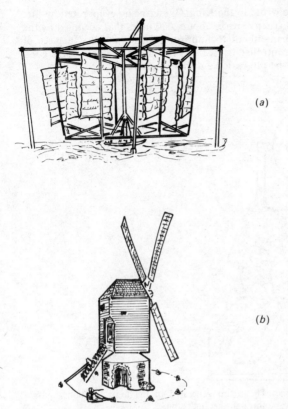

Fig. 1.4. (*a*) The horizontal windmill—still in use in China. (*b*) The post-mill was common in many parts of eighteenth-century Europe.

vacuum', a principle that had been demonstrated in a practical fashion by De Caus (1576–1626), who was a designer of garden waterworks. He connected an almost empty closed vessel of water to a well. He then lit a fire under the vessel and when the water had boiled away he took the fire away, closed the steam vent and the water in the well was sucked up to fill the empty space caused by the condensation of the steam.

De Caus, Galileo (1564–1642) and Torricelli (one of his pupils) were intrigued by the vacuum and the properties of steam, and in the seventeenth century they had established a great amount of scientific information concerning steam. They knew that when water is boiled it turns into steam, and that the volume of steam produced is much greater than the original volume of water (1700 times greater). They knew from experiment that there was great power in the steam. They knew that the steam could be condensed to its original form by cooling, and that if this process took place in a closed vessel a high degree of vacuum could be obtained.

The technological significance of these scientific discoveries was appreciated by Sir Samuel Morland, who was Master Mechanic to King

Charles II. In 1685 he wrote to the King: 'Water being evaporated by fire, the vapour requires a greater space, about 2000 times that occupied by the Water. And rather than submit to imprisonment it will burst a piece of ordnance. But being controlled . . . it bears its burden peaceably, like good horses, and thus may be of great use to mankind . . .'

Fig. 1.5. Hero's Aeolipile. The steam generated in the lower container passed upwards to escape through the angled pipes.

Hero of Alexandria demonstrated this controlled power of steam many years before when he developed, for entertainment, his aeolipile (Fig. 1.5). But it was nearly two thousand years later before steam engines, as we know them today, were envisaged. Denis Papin, a Huguenot working in Germany, devised a means of using a steam-created vacuum to do work. In 1690 Papin placed some water in the bottom of a vertical tube, closed at one end and fitted with a piston. When the water was boiled the piston was driven upwards by the steam pressure, and at the top of its stroke the piston was held by a catch. Papin then cooled the tube, the steam condensed and caused a vacuum and when the catch was released the piston was sucked back into the tube. This was to be the basis of operation of our modern steam engines, but unfortunately financial problems did not allow Papin to pursue it further, and he was to die in poverty.

Captain Thomas Savery was the first man to make a useful steam engine. His engine (1698), known as 'the miner's friend', was used for pumping water out of mines. It used two vessels, alternately filled with steam to drive water out, and then cooled to suck more water in. Since the engine used atmospheric pressure to drive the water into the vessels, it was known as an 'atmospheric engine'.

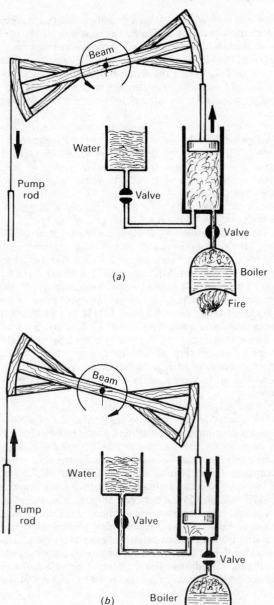

Fig. 1.6. Newcomen's engine: (*a*) the weight of the pump rod pulls the piston upwards, drawing in the steam; (*b*) at the top of the stroke the steam valve is closed and the water valve opened.

Savery's engine had many drawbacks. It was dreadfully inefficient, only putting to good use about 1 per cent of the energy available in the coal that was used in the boilers. And in spite of their gargantuan proportions the early 'miner's friends' could only develop about 1 kilowatt. Another disadvantage lay in the fact that since the steam was produced at atmospheric pressure it could raise the mine water through only about 10 metres. Savery was never really successful in his attempt to overcome this by using high-pressure steam.

Thomas Newcomen, a contemporary of Savery, produced a more efficient engine in 1712, adopting the cylinder and piston method suggested by Papin. Newcomen was aware of the constructional problems involved in the use of steam at high pressures so he restricted his engine to atmospheric pressure. The essential parts of his engine were a boiler and, vertically above it, a piston attached to a pump handle (Fig. 1.6). When the steam was introduced to the bottom of the cylinder the weight of the pump rod raised the piston. When it reached the top a fine jet of cold water was fed in through an injection cock, condensing the steam. The pressure of the atmosphere then forced the piston downwards and so the cycle continued.

Like 'the miner's friend', Newcomen's engines were enormous. The cylinder of one engine at Edmonston, Midlothian (1725), had a diameter of 74 centimetres and a stroke of 2.74 metres; one in Newcastle (1765) had a diameter of 1.88 metres and a stroke of 3.2 metres. Details of one of the earliest engines at Dudley Castle, Worcestershire (1712), tell us that the beam made 12 strokes a minute, generating about 4 kilowatts. By 1769 there were about a hundred of these vast engines in operation across the country. But however magnificent they were in appearance, these engines were terribly inefficient and consumed enormous amounts of coal.

It was up to James Watt to hit upon a method of improving the performance of the steam engine.

Watt was an instrument maker at Glasgow University and his interest in steam power was aroused when he was asked to repair a model of a Newcomen engine. The steam consumption of the model was excessively high and Watt was quick to realise that this was caused by the need to heat up the cylinder each time the steam had been condensed by the cold jet of water. Steam was being wasted in heating up the cold metal. A flash of inspiration came to Watt whilst out walking on Glasgow Green one fine Sunday afternoon in 1765—he would not condense the steam in the actual working cylinder, but would remove it to another cylinder for this purpose. This other cylinder was to be known as the condenser, and its use meant that the working cylinder did not need to be cooled once every cycle. The efficiency of the steam engine was thereby greatly improved (to a magnificent 5 per cent), and this turned steam into a very desirable form of power—power which led to the industrial revolution (see Chapter 10).

1.5 Science and Technology

Now it was at this stage of the evolution of heat engines that science began to assist technological developments, and this interaction grew with great rapidity until today when nearly every technological advance has had to depend on a scientific advance. But in spite of this there is now and always has been a slight edginess in the relationship between the scientist and the technologist. The scientist searches for knowledge and wants to answer the question 'Why?'. The technologist wishes to apply knowledge in his quest to answer the question 'How?'. And in the modern age the required knowledge is usually scientific: for example, the technologist cannot build a nuclear reactor without reference to the scientist's storeroom of knowledge of the phenomenon of nuclear fission. But the technologist has often felt that the contribution of science to technology has been exaggerated, and particularly in the early days when it was claimed that the steam engine was one of the noblest gifts of science to mankind. This view was repudiated by R. A. Meikleham in 1824 when he said: 'There is no machine or mechanism in which the little the theorists have done is more useless. It arose, was improved and perfected by working mechanics—and by them only.'

This is typical of many debates since. It has some foundation: indeed Newcomen had no scientific training or connections. But it must be remembered that the scientists Galileo, Torricelli and Papin had long before established the concept of the vacuum and how it might be used to perform work. So the germ of the idea came from science, and technology carried it through to a practical end.

In many of today's technological advances, the role of science is much more obvious. The scientific understanding of the chemistry of gunpowder has led to developments from cannons to space rockets. The scientists' understanding of the properties of the lodestone led to the mariner's compass. The scientists' study of the life of the silk worm led to the development of silk. But we should not forget that much of the science is actually engineering science; that is, science pursued by engineers. This is amply illustrated by the development of the aeroplane where much of the basic scientific knowledge relating to aerodynamics was discovered by engineers.

The sensitivity of the technologist to the claims of the scientist has its roots in antiquity. It was Plato who encouraged and admired the thinking man, but spurned the practical arts. And this idea was promulgated in Britain for many years, so that even today the engineer is still concerned about his image in the public eye.

Nevertheless, many of today's industries are science-based, but this in no way detracts from the challenge and importance of technology. An enormous amount of scientific knowledge was needed to send probes to the planets, including astronomy, mathematics, physics and chemistry. But at the end of the day it was the technologist who put all this knowledge together in a useful way.

In spite of the growth of science there is still a considerable body of empirical technology that must not be forgotten. Empirical knowledge is knowledge based on observation and experiment and not on theory, and it played an important role, particularly in the early days of technology. The development of weapons, for example, was largely empirical until gunpowder came on the scene. Even today, the design and structure of many musical instruments has been based largely on experience. Again, the development of the printing press was little affected by the scientific revolution. The same could be said of most mechanisms and machines.

And the whole field of materials, from glass manufacture to metalworking, has only recently been assisted by science. Technology and applied science are different things, and it is important to note that technologies can be far advanced by empirical methods before there is a clear understanding of the basic principles on which they depend.

The relationship between science and technology is a symbiotic one, each contributing to and relying upon the other: this is aptly illustrated if we revert to the history of heat engines. We argued earlier that scientific observation of the vacuum must have influenced the technologists who developed the first steam engines. And those engines in turn aroused the curiosity of scientists, not only about engines but about energy in general. The science of **thermodynamics** was born, culminating in the statement of its two famous laws which, in short, are: first, energy can neither be created nor destroyed; secondly, heat cannot of its own accord flow from a cold to a hot body.

1.6 The Conservation of Energy

The First Law, the law of conservation, required a great leap in Man's imagination and was largely due to the work of Mayer (1814–87), Joule (1818–89) and Helmholtz (1821–94). We now realise that energy, defined as the capacity to do work, exists in many forms.

Perhaps the most familiar is **kinetic energy**, the energy of motion, and we know from experience that kinetic energy can do work—a cannon ball can knock a wall down.

Then there is **potential energy** or stored energy. Potential energy is often created by gravity: we know that when the hammer of a pile driver is raised its stored energy is capable of doing work when released. But there are other forms of potential energy, such as that stored in a stretched spring, called strain energy by engineers.

Chemical energy is the energy of food and fuels. We know that the chemical energy of gunpowder can do a lot of work. With the chemical energy available in food, we can manage to do a good day's work, and the chemical energy available in petrol, gas, wood and coal can drive our engines.

Electrical energy is one of the cleanest and most convenient of energy sources. It uses the energy involved with magnets and electric currents. It lights our lamps, our electric fires and drives the ubiquitous electric motor.

Heat energy is used in every steam engine and in every internal combustion engine. Gases expand when heated, and this expansion is put to good use in driving pistons and turning cranks. Heat energy is an extremely subtle form of energy and we shall have more to say about it later.

Sound energy is transmitted by means of periodic pressure fluctuations. Since the main task of sound is to cause movements of our eardrums, the energies involved are normally small (except in a disco!). There are, however, applications requiring greater levels of energy, such as the use of high-frequency sound waves (ultrasonics) to clean dirty materials.

Light energy permeates the whole universe. Its effects are perhaps not so obvious as for the other forms of energy. Yet all of our green vegetation depends upon light energy from the sun, and we in turn depend upon the plants for sustenance.

Nuclear energy is generated within the nucleus of the atom. Some nuclei can be split (fission) causing a chain reaction and releasing vast amounts of heat and mechanical energy. Other nuclei can be combined (fusion) to release energy.

The above list is not exhaustive but it includes the most important forms of energy known to Man today. It allows us to pursue further the significance of the First Law of Thermodynamics. Energy is conserved; it is neither created nor destroyed, *but* it can be transformed from one form to another.

Let us consider, for example, the generation of electricity from coal (Fig. 1.7). First the coal is burned to convert its chemical energy to heat energy. This is then used to drive steam turbines where it is turned into energy of motion (kinetic energy). The turbines turn the huge generators that develop electrical energy and this is then distributed to our homes where we may use it to heat the room (heat energy), light the room (light energy), or drive the vacuum cleaner (kinetic energy). So there are many different kinds of energy transformation in this process.

But in any process involving energy, the First Law tells us that the books must balance, and that the total energy going into a system must equal the total energy coming out, no matter what form it takes. To take a single example, let us consider a rock falling off a cliff. Initially the energy of the rock is all potential energy. If it falls to the ground what has happened to this energy? According to the First Law energy cannot be destroyed, but it may be transformed. In this particular case the friction of the air and the deformation on impact of the stone and the ground generate heat. The original potential energy is all transformed to heat energy, but none has been lost.

The Subtlety of Heat

This is an appropriate point to take a closer look at heat, that subtle form of energy that the earliest scientists found so difficult to understand. It was first envisaged as some form of fluid, the caloric, that could flow from hot to cold bodies. But it is now known that heat is motion at the molecular level, and the hotter a body is the faster its molecules dash about. This **kinetic theory of heat** explains all the phenomena we associate with heat. For example, when a closed container of gas is heated, the molecules hit the walls with greater force causing a rise in pressure. It also helps us to understand the concept of absolute temperature, for the zero on the absolute temperature scale (0 Kelvin or $-273°$ Celsius) is the point at which all molecular motion ceases.

But what is the subtlety of heat? It is basically that all forms of energy will ultimately transform to heat. Heat is the energy graveyard where all other forms of energy will find themselves one day.

You drive to work, for example: where has the chemical energy in the petrol gone to? It has all been converted to heat—in the radiator, in the

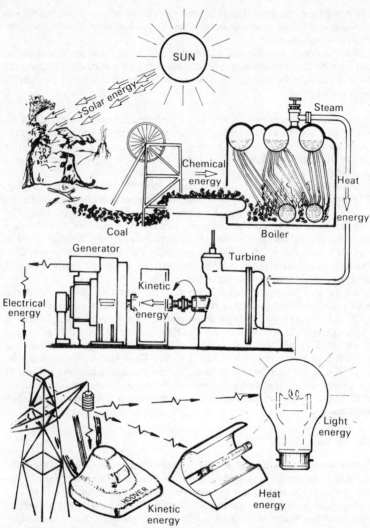

Fig. 1.7. Energy transformations.

exhaust, in the engine block, in the tyres, in the road, in the bearings, and even in the air. And although the original energy has not been destroyed but transformed, it is now practically useless. Engineers say that it is less available. In fact, although we can convert all forms of energy into heat, it is impossible to reconvert heat completely. This is basically due to the 'disordered' nature of heat energy. It is the result of countless numbers of molecules dashing about in random directions and although we can increase their speeds, we do not know how to make them organise themselves!

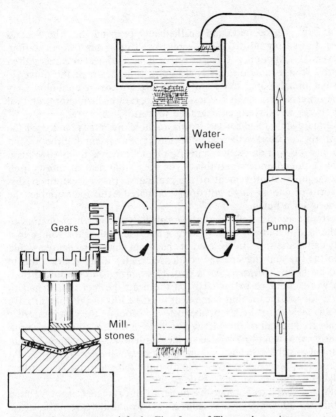

Fig. 1.8. An attempt to defy the First Law of Thermodynamics.

The First Law of Thermodynamics was a magnificent triumph of science and is of immense value to the technologist, whether it be in the design of nuclear power stations, of rockets, of engines or of air-conditioning systems. But there have been instances of people, including engineers and scientists, trying to beat the First Law, albeit unwittingly. Think of the crafty miller who decided that he did not need a river to power his waterwheel. All he needed was a reservoir of water above the wheel (Fig. 1.8). As the water fell from this reservoir it turned the wheel and the wheel did work on the grindstone and drove a pump that sent the water back up to the reservoir. He clearly did not know his First Law. If he had, he would have realised that, in falling, some of the potential energy of the water is converted to heat in the churning water and in the wheel's bearings. Thus, even with a perfect frictionless pump, there would not be enough kinetic energy in the wheel to drive the water back to its original level. The miller had not balanced his books properly and had not included heat energy in his sums. And to suggest that there would still be sufficient energy left over to drive the millstone was stretching it a bit far!

Entropy

The First Law is profound, and aesthetically pleasing, but the Second Law—heat flows from hot to cold—seems to be trite and not worth stating. It is an everyday experience. But this one-way flow of heat was recognised by the early scientists as ultimately leading to the death of the universe. Heat always spreads out to lower and lower levels of energy until ultimately the whole universe will be at a constant temperature throughout, and all forms of energy will have been degraded to heat.

This ultimate state of 'disorder' is unavoidable. Scientists have used the word 'entropy' to describe the state of disorder of a system and the Second Law can also be written as: the entropy of the universe tends towards a maximum. You can understand this if you recall that heat manifests itself as a chaotic jumbled movement of the molecules. It is an extremely disordered form of energy (high entropy), compared with, for example, the kinetic energy of a bullet.

The Second Law also tells us that we can only get our heat engines to work if the heat is allowed to flow from a hot body called the source to a cold body called the sink. In the case of the steam engine the source would have been the fire and the sink the condenser. Sadi Carnot realised this in 1824, and his book *Réflexions sur la puissance mortice du feu* showed that, although work could be extracted from the heat, the fact that it had to flow to a cold sink meant that some heat must be lost in the cold sink. In practice, this meant that heat must be lost when cooling the steam in order to condense it. Thus all of the energy available in the fire could not be put to good use; some would be wasted in the condenser. Carnot showed that the maximum possible **efficiency** of a heat engine was limited to $1 - T_2/T_1$, where T_2 and T_1 were respectively the absolute temperatures (see earlier) of the cold end and the hot end of the engine.

This was to give the technologists a target to aim at. The Carnot formula showed that the efficiency of the steam engine would increase if the steam temperature T_1 could be increased. Now as we all know, no matter how long we boil a kettle the temperature of the steam will not rise beyond 100°C. But the steam temperature can be increased beyond 100°C if we use a pressure cooker, for it is a well-known fact that the boiling point of water increases with pressure. So the engineers realised that in order to increase T_1 they would have to increase the steam pressure. Nowadays steam turbines operate at pressures exceeding 160 atmospheres, and at temperatures around 500°C.

What is the maximum achievable efficiency of a practical engine? Looking at a modern engine the upper limit on T_1 would be about 873 K (600°C + 273), since above this temperature steel begins to lose its integrity. Taking T_2 as 15°C (288 K), the temperature of the atmosphere into which the engine exhausts, gives an efficiency of $1 - 288/873 = 67$ per cent. Newcomen's 1 per cent efficient engines fell far short of this ideal.

We have now traced the development of Man's use of energy up to a period just preceding the industrial revolution. Later we shall have a lot more to say about energy, including the quest for new energy sources and the growing need for a fresh look at the older sources such as windmills and waterwheels. But, quite a few pages back, it was proposed to examine two areas where the growth of Man's knowledge illustrated an evolution from empiricism to science-based technology. One was energy; the other was concerned with Man's use of materials—and we will now look for evidence of a similar evolutionary growth in that field.

1.7 Materials

Technology is largely concerned with processes of transformation; that is, changing raw materials into useful and pleasing things. What raw materials did early Man have at his disposal? There were the four basic elements of the Greeks: fire, air, water and earth—and we have just seen how the first three were exploited as energy sources. But the last basic element, earth, was to provide an abundant supply of materials that Man could use to solve his problems. The use of materials was of the utmost importance to Man, for without strong durable materials, tool-making could not have developed, and without tools technology cannot advance. Thus, throughout history the evolution of technology has been largely determined by the materials available.

Earth provided Man with a prodigious amount of material, both organic and inorganic. There were, for example, grass, leather, fur, ivory, wood, and many other organic substances that could be used for clothing, weapons and tools. But we shall have to restrict ourselves to a bird's eye view of the developing use of inorganic materials, in particular stone and metals. These materials played such an important part in the growth of civilisation that historians have designated three great ages in Man's history: the **Stone Age**, the **Bronze Age** and the **Iron Age**.

The Stone Age

The most obvious naturally occurring material is stone, and it is not surprising that Man was quick to appreciate its hardness, durability and availability. The Stone Age can be subdivided into three important eras; the **Paeleolithic** or the Old Stone Age, the **Mesolithic** or Middle Stone Age, and the **Neolithic** or New Stone Age. The Paeleolithic Age was by far the longest period, covering most of human history and starting with the dawn of mankind about two million years ago. The Mesolithic Age covered the period between the last of the Ice Ages up to the time when Man ceased his wanderings after animal prey and settled down to an elementary form of domestication. Its length varied from area to area, but in Europe the Mesolithic Age lasted about 5000 years, from the ninth to the fourth millenium BC.

It has been estimated that emergent Man used stones in their natural form as tools more than two million years ago, and there is evidence that Man's ancestors, the **hominids**, in particular *Australopithecus* (the Southern Ape Man), actually manufactured tools from stones as long as one and a half million years ago. These were simple pebble tools, made from pebbles of lava or quartz about the size of a tennis ball. They were manufactured by using another stone as a hammer to chip off small pieces from one end of the pebble to form a cutting edge. It is hard to appreciate that this simple tool was to be employed, practically unchanged, for a period at least fifty times as long as recorded history. It was used for skinning and butchering and for digging up bulbs and roots from the earth.

As time went by the advantages of flint made it popular for tool making. Flint is very common in Europe, where it is found in nodule form in chalk deposits. It is black or brown in colour and the nodules, which are often washed out by the rivers, are covered with a rough brown or white crust. But most important of all, flint can be worked to give a very sharp edge for, when struck, it suffers a conchoidal (shell-like) fracture that is curved and has a concentrically ribbed surface. Another mineral, obsidian, was often used, especially in the New World where it was much more abundant than flint. Obsidian, which looks like black glass, could be worked to a fine edge.

A flint could be made into what is called a core implement by using a stone (the hammerstone) to chip bits off it. Alternatively, each chip could be made into a separate tool, a flake implement, and these were often used for scraping the fat off the inside of animal skins before they were used for clothing or tenting. Hand axes were very popular, the earliest ones being rough core implements made with a hammerstone; but as the technology of tool-making developed it was discovered that a hammerstone was not necessary and flakes of flint could be removed if you pressed hard enough with a bone or wooden punch. Axes made by this pressure flaking method were much finer in appearance and effectiveness. Towards the end of the Paeleolithic Age, 75 000 years ago, Neanderthal man was equipped with complete tool kits, made from flakes, and including saws, knives, scrapers and borers.

When the Ice Ages ended about 10 000 years ago, the Mesolithic Age commenced and at that time modern Man, or *Homo sapiens sapiens*, had already been in existence for 25 000 years. The change in climate caused great changes in his mode of living. He lived in villages and ate plants and fish as well as meat, and, in order to catch his prey in the new thick forests, he developed the bow and its flint-tipped arrow.

The last major technological advance in the manufacture of stone tools was the discovery that stone could be polished, using sand and water, to give a sharp edge. This heralded the arrival of the Neolithic Age when Man emerged from savagery. It was a period of rapid growth in population, and in the widespread introduction of agriculture, particularly in lands to the south-east of Europe where climate and geographic conditions were ideal for the growth of civilisation. And the Neolithic revolution was to lead to Man's discovery of metals and metalworking.

Copper

That part of the world now comprising northern Iraq, north-western Iran and eastern Turkey saw the beginnings of metals, and dress ornaments were known to have been hammered from naturally occurring copper as far back as 8000 BC. In addition to copper there were several other naturally occurring or native metals that Man discovered lying about on the surface of the earth and in mines. These included gold and silver, which were greatly prized and used for decorative purposes.

And meteoric iron, being very rare, was greatly valued for tool-making.

These metals could be worked in two ways. They could be hammered into shape, and it was soon discovered that this process tended to harden the metal and that, if necessary, the metal could be softened again by heating. The other method of shaping the metal was to melt it and pour it into moulds.

It became necessary to extract copper from its ore when the supplies of the native metal had dwindled, and about 4000 BC most of the copper in the Middle East was obtained by smelting. This discovery of smelting probably resulted from Man's fascination with fire. It was an object of veneration and a symbol of unspeakable mysteries. In many metallic ores the desired metal exists in chemical combination with sulphur or oxygen, and it was a great advance for Man when he discovered that carbon could be used as an alternative partner for these elements, thereby leaving the metal in its pure form. And it was a bonus that the carbon could also be burned to provide the heat for the reaction. The smelting process was complicated and records show that in the eastern Alps around 1000 BC copper ore and charcoal had to be recycled three times in the furnace to produce 95 per cent pure copper.

Copper produced in this way was a relatively soft metal, and it was a surprise when Man discovered around 3000 BC that copper became stronger when it was less pure and included contaminants like tin, antimony, arsenic or zinc. When copper is mixed with one or more of these elements the resulting alloy is known as **bronze**.

The copper-tin alloys in particular were most useful and, in spite of the scarcity of tin, they dominated metallurgy for nearly two millennia. Bronze was harder than copper and was more easily worked. Its melting point was lower than that of copper, and the smiths found, using the same temperatures at which they had melted copper, that they had a much more fluid metal. The quality of casting improved dramatically. Complicated moulds of two or more pieces were used to produce ornaments, tools and weapons. Great detail was achieved using the 'lost wax' process. When applying this technique the artisans first produced a wax model of the final product. This was then surrounded by clay and baked in an oven, and during the baking the wax melted and ran away, leaving a clay mould into which the molten bronze would ultimately be poured. These techniques produced accurate and extremely durable artifacts, and indeed many of the craft tools that we know today were developed during the **Bronze Age**, e.g. the sledgehammer, the cold chisel, the rasp.

The early metallurgists experimented to determine the effects of varying the proportions of copper and tin in their bronzes. For example, they found that in the manufacture of mirrors it was necessary to use a lot of tin to produce the desired whiteness in the metal. But this resulted in a brittle metal that was useless for tools.

Iron

The next advance in metallurgy was the large-scale extraction of iron from its ore, heralding the start of the **Iron Age** about 2000 BC. The Hittites, who lived in what is now the eastern half of Turkey, were the first people to produce a steady supply of iron, and it seems that it was the economic rather than the inherent qualities of the metal that gave rise to the widespread use of iron. Iron ore was abundant; certainly much more so than tin. Its manufacture required only the simplest of equipment, and wherever there was iron ore and wood, iron could be made if you had the 'know-how' (Fig. 1.9).

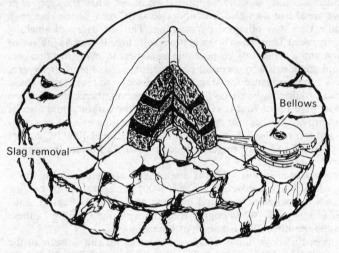

Fig. 1.9. Smelting iron. The primitive blowing hearth consisted of alternate layers of charcoal and iron ore on a stone platform (*ca.* 1400 BC).

But the early irons were no harder than cold-worked bronze; and they were difficult to work to a fine finish. In those early days it was not possible to generate sufficient heat in the furnaces to melt iron. Copper melts at 1083°C, but pure iron will not melt until the temperature is raised to 1535°C, and this was not possible until relatively recent times.

Thus the early metallurgy of iron was quite different from that of copper. Since it could not be melted, the spongy iron that resulted from the reduction of the ore with charcoal had to be consolidated by hammering when hot. This hammering produced bars of wrought iron from which more complicated forms could be produced by forging and welding.

The techniques of iron-working developed by a long process of trial and error. There were several surprises. For example, the early natural philosophers and smiths had thought that steel was a purer form of iron rather than a contaminated form, and it was a long time (AD 1774) before it was realised that steel was actually an **alloy** consisting of iron and carbon. This is understandable since the alloys of copper had been manufactured by consciously mixing something with the copper, but the contaminant required for steel was introduced to the iron accidentally in the process of heating it.

Another surprise was that steel could be further hardened by quenching the hot metal in cold water. If this had been done to the copper or bronze, they would have become softer. But the metallurgical principles behind quenching were not properly understood, and the high quality of steel that was produced occasionally was highly prized and sometimes considered to have magical properties—perhaps giving rise to the legendary Arthur's Excalibur and Siegfried's Balmung of later times.

Because of an inability to generate a sufficiently high temperature, the low-temperature reduction of the iron ore with charcoal remained the principal method of manufacture of iron for about 3000 years. Furnaces increased in size during the Middle Ages, and forced blasts of air from bellows were used to increase temperatures. These blast furnaces were ultimately able to produce molten iron (Fig. 1.10). In the fourteenth century iron was being poured from the furnaces of the Rhineland, and by the end of the sixteenth century the technique was widespread.

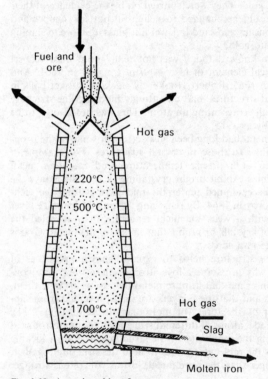

Fig. 1.10. A modern blast furnace.

Science of Materials

In our brief look at Man's exploitation of energy sources, we saw that science first began to make a substantial contribution to technological developments in medieval times. The story is much the same for materials, although on the whole the most important scientific advances did not come until more recent times.

Two aspects of the science of materials can be identified. One, the microscopic, is concerned with the molecular structure of materials. The other, the macroscopic, relates to bulk properties such as resistance to tension and compression, the conduction of electricity and of heat.

The study of microscopic aspects of materials was founded in the science of **chemistry**, which was not really recognised as such until the close of the Iron Age. Chemistry developed from the technology of metal workers, jewellers and potters who had developed by trial and error, the skills of smelting ores, purifying and colouring metals and pots.

In addition to the quest for the elixir of life, the ancient alchemists were fascinated by **transmutation**—the changing of substances into other substances, especially into gold. They were spurred on by the evidence of their eyes: clay and water could be changed to delightful pottery; copper and zinc to beautiful brass; ashes and sand to jewel-like glass; dull ore to shining silver. Why not lead into gold?

But chemistry as we know it today was not really born until Robert Boyle defined a chemical element in his *Sceptical Chymist* in 1661. And just a few years later in 1665, Robert Hooke, by stacking musket balls in piles to simulate crystal structures, had unwittingly modelled the structure of metals, with each ball representing an atom. This was not known to be the case until as recently as 1915.

The **crystallinity** of metal had long been suspected, for some of the properties of metal were alike to those of crystal structures. For example, a molten metal solidified at a precise temperature, and fractured metal surfaces would sometimes exhibit bright crystalline facets. Nowadays the crystal structure can be examined under the microscope using the techniques developed by Sorby in 1864. By removing the polished surface layer of metal by etching with a weak chemical reagent, Sorby revealed the irregular honeycomb of crystals or grains that make up the metal. Grains are typically about 0.25 mm across.

A knowledge of the structure helps to explain known properties of metals. For example: why are some alloys stronger than their pure constituents? Well, one can argue that in pure metal the atoms arrange themselves in orderly layers and that these layers are only prevented from slipping over one another by the forces of molecular attraction (Fig. 1.11). But when an alloy is made, atoms of different size and shape are introduced and these, by disturbing the regularity of the structure, provide a locking mechanism that assists the molecular forces to hold the structure together. Thus the alloy is stronger than the pure metal, just as concrete is stronger than cement.

Nowadays, as well as strength, engineers are concerned with the electrical, magnetic, thermal, nuclear and acoustic properties of materials. In addition, cost, weight, durability and appearance are of great importance. With each new design, the engineer has to compromise among these properties and the scientist has to tell him how to achieve the required structure. Thus the scientific understanding of materials allows chemical 'tailoring' to produce materials for specific purposes. And today these materials need not be metallic, for science and technology have together perfected other useful materials like ceramics and plastics.

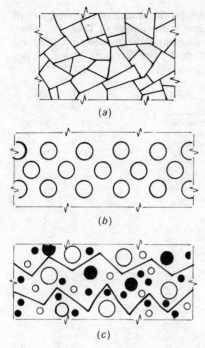

Fig. 1.11. (*a*) Mass of small crystals or grains that make up a metal. (*b*) Orderly lattice structure of atoms in a pure metal. (*c*) Interlocking within an alloy.

1.8   Technology Today

Modern technology covers a wide spectrum. It is subdivided into many compartments, each consisting of its own specific developed body of knowledge. One way of categorising them follows:

Biological technology	agriculture, forestry, fishing
Environmental technology	building, housing, lighting, heating and ventilation, acoustics
Mining technology	ores, oil, coal, gas
Transport technology	road, rail, air, sea, space
Chemical technology	chemicals, polymers, metals
Communications technology	radio, telegraphy, television, printing
Materials technology	plastics, glass, metals, ceramics, rubber
Energy technology	gas, steam-raising, electricity, nuclear, alternatives
Food technology	baking, preservatives, distilling, brewing
Medical technology	orthodontics, prosthetics, life preserving, medical aids.

Another method of categorising relates the branch of technology to its most important knowledge base. This is summarised in Table 1.1, which shows some of the many overlaps amongst the technologies. For example, food technology requires an input from chemistry as well as biology.

Table 1.1

Knowledge base	Branch of technology
Biology	Fishery
	Forestry
	Agriculture
	Food
	Food
	Oil, gas, coal
Chemistry	Chemicals
	Adhesives
	Materials
	Materials
	Vacuum
Physics	Acoustics
	Nuclear
	Electronics
	Electronics
	Power generation
Electrical engineering	Computing
	Communications
	Medical
	Medical
	Energy
Mechanical engineering	Machinery
	Manufacture
	Transport
	Transport
	Building
Civil engineering	Housing
	Water supply
	Drainage

It would be extremely difficult, and probably not worthwhile, to draw up a table showing all of the technologies and all of their possible interactions. Neither of our methods of categorisation is comprehensively accurate, but both should give you a clearer idea of the breadth of technology. You should also be able to appreciate that many technological projects require the combined efforts of technologists from many different fields. Building an aircraft, for example, needs people with expertise in engines, materials, communications, transport, acoustics, etc.

Table 1.1 also raises another important point: the relationship between **engineering** and technology. A technologist is someone who practises technology, so although all engineers are technologists, a technologist need not necessarily be an engineer. For example, an electrical engineer is a technologist, but a food technologist is not an engineer. There are many technologists in addition to the engineers—building technologists, quantity surveyors, food technologists, environmental health officers, computer technologists, to name but a few. I make this point for technology and engineering are often taken to be synonymous. This is wrong. Engineering is a branch of technology.

1.9 Concluding Remarks

This chapter started by defining technology as the systematic application of scientific or other organised knowledge to practical tasks. The 'systematic application' is the methodology we referred to earlier, starting from the recognition of a need, right through to the manufacture of an artifact or system to meet that need. We shall devote more time to this problem-solving activity later in Chapter 7.

With particular reference to Man's growing use of energy and materials we have been able to compare the relative importance of 'scientific and other organised knowledge'. It was demonstrated that the empirical approach was predominant in the earliest stages of the growth of technology, and that the systematic development of technical inventions from scientific discoveries did not really become conspicious until about the nineteenth century. Nowadays the relationship between science and technology is a reciprocal one, and advances in science are soon applied to new technologies, and vice versa. This is nicely illustrated by the interactions between developments in computing and in mathematics.

We also noted that today's technology covers a very wide spectrum of disciplines, and we made an attempt to categorise these.

1.10 Exercises

1. Describe and compare the methodologies of science and of technology. Give four examples of scientific contributions to technology.

2. Distinguish between empirical knowledge and scientific knowledge. Give two examples of each and explain how they contributed to the advance of technology.

3. Sketch four primitive agricultural implements. List the major milestones in the development of agriculture.

4. Define the term 'energy'. List six forms in which energy exists, and give two examples where each form can be put to good use.

5. (*a*) Describe two technological innovations which had a beneficial effect on the utilisation of water power. (*b*) What technological advance was mostly responsible for the decline of the water mill?

6. Name two civil engineers, two mechanical engineers and two electrical engineers who made important contributions to engineering. Describe their contributions.

7. List three important properties of each of the materials metal, wood, plastic. Indicate how the strength of each of these may be increased.

8. List the advantages and disadvantages of using (i) plywood, (ii) mild steel, (iii) glass-reinforced plastic for the body of a motorcar.

9. What personal attributes do you think essential to a successful technologist? Would your list be the same for a scientist?

10. Write short essays (100 words) on the job of a food technologist and of a chemical technologist.

1.11   Further Reading

Those marked with an asterisk are particularly recommended.

Angrist, S. W., and Helper, L. G., *Order and Chaos*, Basic Books, 1967.

Bernal, J. D., *The Extension of Man*, Weidenfeld and Nicolson, 1972.

Bernal, J. D., *Science in History*, Penguin Books, 1965.

* Bronowski, Barry, Fisher, and Huxley, *Technology, Man remakes his world*, Macdonald, 1973.

Cardwell, D. S. L., *Technology, Science and History*, Heinemann, 1973.

* Derry, T. K., and Williams, T. I., *A Short History of Technology*, Oxford University Press, 1960.

Energy, Time-Life Books, 1969.

Energy Conversion: Power and Society, Open University Press, 1972 (T100 Units 20–21).

Engineering Heritage, Heinemann Educational Books, 1966.

Fundamental Concepts in Technology I, Unit 4 Energy, Open University Press, 1975.

Holliday, L., *The Integration of the Technologies*, Hutchinson, 1966.

Machines, Time-Life Books, 1969.

Millett R., *Design and Technology of Plastics*. Pergamon Press, 1977.

Rogers, C. F. C., *The Nature of Engineering*, Macmillan, 1983.

Shackley, M., *Rocks and Man*, George Allen and Unwin, 1977.

Smith, C. S., 'Materials', *Scientific American*, September 1967, Vol. 217, No. 3.

2
Technology Extends Man's Muscles

2.1 Man's Limitations

Although Man is an extremely versatile creature, his size, his structure and
his muscles put severe limitations on his capabilities. These physiological
limitations exclude Man from many tasks but invariably when such situations arise technology has been brought into play. The extension of Man's
capabilities is a primary role of technology. Referring to Fig. 2.1, we can
see that Man's muscle power has been extended by machines and mechanisation. His brain has been augmented by the computer. His ability to
measure, indeed his senses, have been extended by all sorts of instruments
and measurement devices. As a controller of an operation, we have many

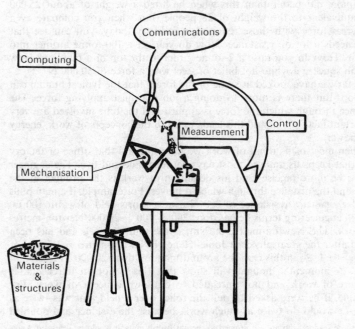

Fig. 2.1. Technology extending Man's capabilities.

shortcomings including our inability to concentrate for long periods and our inbuilt delay in response to input signals. (It takes the average human about 0.2 seconds to react to a stimulus.) Developments in control systems now allow us to control operations with great accuracy and speed. Technology has also allowed us to communicate faster and further than our voices alone would ever have allowed. And finally we must not forget that the materials and structures for our new artifacts depend a lot on technology.

We shall meet most of these areas of human capability in this and the following three chapters, but first let us take a quick look at how **mechanisation** has extended muscle power. It is not easy to state the maximum force that a person can develop, for it depends on so many things: the way in which the force is applied (pull or push?); the variation from person to person; the variation from situation to situation. For example, if we consider the forces that can be exerted from a sitting position, as shown in Fig. 2.2(*a*), experiments on a group of students have shown that on average an extended arm can exert the following forces: lift, 180 Newtons*; lower, 160 N; push, 550 N; pull, 500 N. These forces will vary when the elbow is bent; in general the lifting and lowering forces will increase, and the pushing and pulling forces will decrease as the angle of the elbow increases. Much greater forces can be generated from a standing position, and similar experiments examining the forces generated by the leg muscles show that on average the students were able to develop a force of about 6000 N in the position shown in Fig. 2.2(*b*). Paul Anderson, the Olympic weight-lifting champion, did better than this when he lifted a weight of about 25 000 N—equivalent to the weight of 38 people. But when you compare even this great force with those required in industry today, you can see that Man needs a lot of assistance. How do you get a 100-tonne airliner into the air? How do you raise a 2-tonne girder to the top of a building? How do you squeeze a white-hot billet of steel with a force of 50 000 N?

So far we have looked at some of the forces that the typical human can develop. But there is more to doing a job than just applying force. The distance through which the force is applied and the time involved are very important factors too, and this leads us on to the concepts of work, energy and power.

When most of us think of work we have in mind that office or factory task that keeps us amused most days between nine and five. The engineer has to be more precise, and his definition of **work** is the product of the force and the distance through which it moves. For example, if a man pulls a roller against a resistance of 200 N over a sports field of width 100 m, then in engineering terms he has done $200 \times 100 = 20\,000$ Newton-metres of work. The Newton-metre, or Nm, is the unit of work and has been called after the great physicist Joule. Hence 1 Newton-metre = 1 Joule, or 1 Nm = 1 J. So in this case the groundsman has done 20 000 J or 20 kJ of work. A moment's thought will show that this is not an unreasonable measure of work and not unrelated to our own concept of work. For example, if he were asked to pull the roller over a field that was twice as wide, he would do twice as much work, because the distance had doubled

---

* For those readers who are accustomed to the old British units, it would be helpful to know that a Newton is approximately one quarter of a pound force.

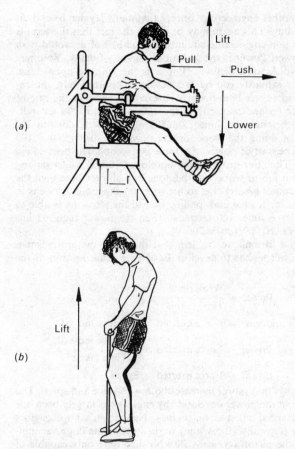

Lift

Pull

Push

Lower

(a)

Lift

(b)

Fig. 2.2. We generate different forces when (a) sitting and (b) standing.

and therefore he would be in a position to claim twice the pay. Even on the same size of field, the work required would increase if the force required to overcome resistance were to increase. This would be the case if a heavier roller were used or if the ground were muddy.

We spent some time on energy in Chapter 1 and it is now important to emphasise that work and energy are related. **Energy** is defined as the capacity to do work. We saw that there were many different forms of energy: potential, kinetic, chemical and so on. The energy possessed by a body, no matter which of these forms it takes, is a measure of the work it is capable of doing. Our groundsman would not be capable of doing his job if his food and his oxygen intake had not given him a total chemical energy in excess of 20 kJ. Turning to another example, a bullet with a kinetic energy of 100 J has the capacity to do work amounting to 100 J. It could do this by burying itself a distance of 10 cm in a material that provided an average resistance of 1000 N ($100 = 1000 \times 0.1$)

Now power is another engineering concept that many laymen have difficulty in understanding. This again may be due to the fact that the term is commonly used in non-engineering circumstances. Most of us would wish to be people of power. Power is the ambitious pursuit of many. You may have heard of Lord Acton's dictum—'Power is the ultimate aphrodisiac but absolute power exhausts you by Wednesday.' Unfortunately, the engineer's definition of power is not nearly so exciting: **power** is the rate of doing work. If our groundsman covers the 100 m pitch in 200 seconds, then according to the engineers, since 20 000 J of work were done in 200 seconds, the rate of doing the work, or the power, is 20 000/200 = 100 Nm/s = 100 J/s. The unit of power, the Joule per second, has been called after James Watt. Thus our groundsman developed 100 W whilst pulling the roller. If a given job of work is to be done in a shorter time then the required power becomes greater. Let us assume that the groundsman is in a rush to get home for his tea and, pulling out all the stops, he is able to do the job in half the time, 100 seconds. Then the power required has increased by two, i.e. 20 000/100 = 200 W.

We can develop a formula to confirm that the faster the groundsman moves the more power he has to develop. Earlier we defined power as the rate of doing work:

Hence $$\text{Power} = \frac{\text{Work done}}{\text{Time taken}}$$

but $$\text{Work done} = \text{Force exerted} \times \text{Distance moved}$$

therefore $$\text{Power} = \text{Force exerted} \times \frac{\text{Distance moved}}{\text{Time taken}}$$

or $$\text{Power} = \text{Force exerted} \times \text{Speed}$$

This formula shows that power increases both with force and speed. The enormous increase in the power developed by engines has largely been due to the increase of speed of modern engines. Newcomen's huge engines turned very slowly (typically 10 rev/min) so, in spite of the large vacuum-induced forces at the piston (typically 20 kN) they were only capable of generating around 4 kW. Today's motorcycle engines are much smaller—compare their piston diameters of around 0.1 m with the beam engine's 1 m. Smaller forces are developed during the power stroke (typically 10 kN) but this, and their small size, are more than offset by their much greater speeds of rotation (typically 7500 rev/min). As a result a small modern engine can easily develop 50 kW of power.

Even after doubling his speed our groundsman is still a long way from generating sufficient power to keep a 1 kilowatt (1000 W) electric fire alight. And he still falls a long way short of the power of a horse, for James Watt estimated that a draught horse could develop about 0.746 kW (originally called the horsepower).

What power can the average human develop? Experiments have shown that for short intervals, say less than one second, a man can develop about 1 kW. But as the length of the working period goes up, the attainable power output reduces. Over a period of about 10 seconds it is possible to maintain an output of about 200 W, but for longer periods this falls to about 100 W. So again, Man's efforts are seen to be puny in comparison to the average car with its 50 kW, not to mention the typical power station with an output of 1 MW. (One megawatt = one million watts = one thousand kilowatts. A full list of such prefixes is given in Appendix 1.)

2.2 Machines

Since the forces and the powers available from a human are relatively low it was necessary in the very early days of civilisation to combine the efforts of many people when big tasks, like the building of the pyramids, had to be done. But a more subtle approach was to use devices to multiply the efforts of a single person. Machines were developed to allow Man to match the small force that he could develop to the larger force required to do the particular task.

'And what,' you ask, 'do you mean by a machine?' It is well nigh impossible to get two technologists to agree on a definition of a machine. This is basically because the concept of the machine has changed so much since the time of the ancient Greeks. In 1876, Reuleaux, the great German engineer, defined a machine as a combination of resistant bodies so arranged that by their means the mechanical forces of nature can be compelled to do work accompanied by certain motions. A more recent version of this defines a machine as a system of force-resistant bodies or machine parts whose motions are constrained, which converts or transmits energy to do work. But you can get the general idea—work is to be done by transmitting energy from one point to another and certain determinate motions and forces are generated at the output. A major source of confusion is the fact that this definition has a distinct bias towards mechanical engineering; and this is not surprising since the first machines were all mechanical by nature. But nowadays many technologists call a computer a machine, and clearly the above mechanically-orientated definition would not be satisfactory. Perhaps a more all embracing definition would be that a **machine** is a device for extending human capability.

However, let us return to Man's early machines, those with a distinct mechanical orientation. In Chapter 1 we introduced the five elementary machines, enumerated by Hero of Alexandria about the time of Christ. They were the lever, the wheel and axle, the pulley, the wedge and the screw.

The Lever

The lever, in the guise of a club, is probably the oldest machine. It consists of a straight bar or rigid structure of which one point, the fulcrum, is fixed, another is connected with the force to be acted on, and a third is connected with the force or power to be applied. Fig. 2.3(*b*) shows a lever in diagrammatic form. In order to understand how it works we must have an appreciation of the moment of a force about a point. The effectiveness of a force to produce rotation is called the **moment** of the force, or its **torque**, and is equal to the product of the force and the perpendicular distance from its line of action to the point about which it rotates. Thus in Fig. 2.3(*b*), the applied force or **effort** E exerts a torque Ee in a clockwise direction, whilst the resistance or **load** L exerts a torque Ll in an anticlockwise direction about the fulcrum. (Note that *L* is the force exerted by the load on the lever; the lever will exert an equal and opposite force on the load.) The system will be in balance (in equilibrum) when the sum of all the torques about the fulcrum is zero—or, in other words, when the anti-clockwise torques equal the clockwise torques. In the case shown we assume there is no friction at the fulcrum so that equilibrum exists when

$$Ee = Ll$$

or when

$$E = L(l/e)$$

Thus a large load can be held by a smaller effort provided that *l* is less than *e*. For example, if *l* were one fifth of *e* then the effort would only have to be one fifth of the load: the small man could balance a man five times his weight and, conversely, the pliers would amplify the grip force five times. The ratio of load to effort (*L/E*) is known as the **mechanical advantage** (MA) of the machine; in this case the mechanical advantage is (*e/l*).

So here we have a device that can multiply Man's puny muscle power by a factor, the mechanical advantage, that can be greater than 1. It's too good to believe and we suspect that there must be a catch somewhere. And

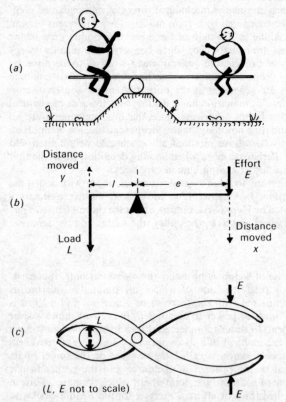

Fig. 2.3. Levers of the first kind: (*a*) a seesaw in (*b*) diagrammatic form; (*c*) pliers.

there is! If you examine Fig. 2.3(*b*) carefully you will note that, in order to move the load *L* through a certain distance *y*, the effort will have to be moved a greater distance *x*; in fact, *e*/*l* times greater. So although in our example we only have to exert an effort equal to one fifth of the load, we will have to apply it over a distance equal to five times the distance through which we wish to move the load. We might have reached this conclusion earlier if we had remembered that Nature does not give anything for nothing—as the Law of the Conservation of Energy reminds us. This can be interpreted in this case as 'work in' equals 'work out'; so if forces increase then displacements will have to decrease to keep the work constant.

In the early days it was common to refer to the ratio (*x*/*y*) as the velocity ratio, but nowadays it is more popular to use the concept of a movement ratio or **transmission factor** defined as:

$$\text{Transmission factor (TF)} = \frac{\text{Distance moved against load}}{\text{Distance moved by effort}} = \frac{y}{x}$$

In the case shown in Fig. 2.3 trigonometry shows that the transmission factor is (*l*/*e*). In general we can show that the mechanical advantage is the inverse of the transmission factor when friction can be neglected.

Archimedes was so carried away with the potential of the lever that he made the wild claim: 'Give me a place to stand upon and I will move the world.' Clearly Archimedes had imagined that if he had an extremely long lever with a very high mechanical advantage, he could have amplified his own muscle power sufficiently to move the Earth. Let's look at this more closely. Let us assume that by 'moving the world' Archimedes meant that he could lift, on the Earth's surface, a weight whose mass was equivalent to that of the Earth. Now the mass of the Earth is about 6×10^{24} kg, so on Earth the weight (see Chapter 3) of such a body would be about 6×10^{25} N. If we assume that Archimedes could exert a force of 1000 N (see the leg lift data given earlier), then the mechanical advantage or the ratio of the arms of the lever would have to be $(6 \times 10^{25})/1000 = 6 \times 10^{22}$. Now let us assume that Archimedes wished to move the Earth by one centimetre. Since the transmission factor is only $1/(6 \times 10^{22})$ then he would have to move his end of the lever by 6×10^{22} cm $= 6 \times 10^{20}$ m. This is an enormous distance—four thousand million times the radius of the Earth's orbit about the Sun. How long would it take him to do this? Even if we make the ridiculous assumption that he could move his end of the lever at the speed of light (3×10^8 m/s) it would still take 2×10^{12} seconds, or about 70 000 years. This is a long time, certainly a lot longer than Archimedes' life-span.

The levers we have just examined have the fulcrum between the effort and the load. They are known as **levers of the first kind**, and include such things as scissors, oars and balances as well as the examples shown in Fig. 2.3. **Levers of the second kind** place the load between the fulcrum and the effort (Fig. 2.4). Their mechanical advantage (*e*/*l*) is always greater than unity, so like the first kind they too amplify the effort. They differ from levers of the first kind, however, in that the load and the effort both move in the same direction.

In **levers of the third kind**, the effort is placed between the load and the fulcrum (Fig. 2.5). Such levers have mechanical advantages less than unity, but their transmission factors exceed unity so that they favour distance and speed at the expense of force. They are used to move objects of small mass with rapidity. Levers of the third kind are common in our own bodies.

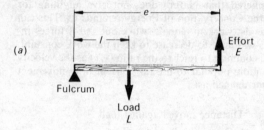

(a)

(b)

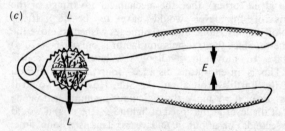

(c)

Fig. 2.4. Levers of the second kind: (*a*) schematic, (*b*) wheelbarrow, (*c*) nutcracker.

It is worth emphasising at this point that machines are used to match men and their tasks. Man can generate sufficient energy to do many jobs, but often the energy is not in the right form. For example, consider how much energy you need to walk up a flight of stairs. You will recall that work is the product of force and distance; in this case the force is your weight and the distance is the vertical height that you climb. So if you weigh 600 N and you climb 2 m, you will use up 1200 J of energy. Now compare this with the energy required to lift one corner of your car sufficiently high to change a tyre. The average car weighs about 10 kN so if you lift one corner through say 0.5 m, you raise the centre of gravity about 0.125 m and the energy requirement is 10 000 × 0.125 = 1250 J. So if you can climb the stairs unaided you ought to be able to lift the corner of the car unaided! But we know from experience that this is not the case. The problem is that the energy developed by us, and that required for lifting the car, take different forms. This is illustrated in Fig. 2.6, where the energy required is shown as the shaded area. (Remember that energy or work is the product of force and distance.) The energy required to lift the car is the product of a large force and a small distance. The energy developed by a human is the product of a small force and a large distance. Although the energies are the same it requires a machine, in this case a jack, to match the man to the task.

The **efficiency** of a machine is another useful concept. It is concerned

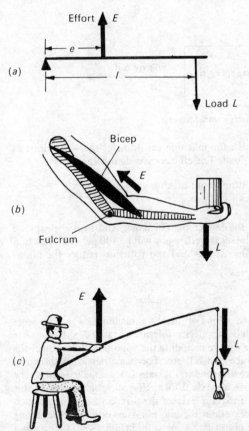

Fig. 2.5. Levers of the third kind: (*a*) schematic, (*b*) human arm, (*c*) fishing rod.

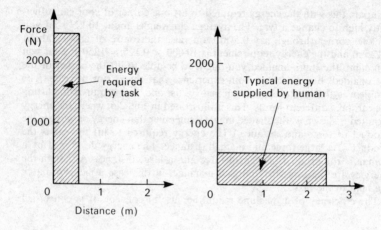

Fig. 2.6. The machine as an energy transformer.

with the efficiency with which the machine can transfer the work done at the input, to the output or load. The efficiency is defined as

$$\text{Efficiency} = \frac{\text{Useful work output from machine}}{\text{Work input to machine}} \times 100 \text{ per cent}$$

The better the machine, the higher the efficiency, and in the ideal case where there are no friction losses the efficiency will be 100 per cent. Friction will reduce the force available at the load and will thus reduce the mechanical advantage.

The Wheel and Axle

The wheel and axle, classified as one of the basic machines, is really a lever that can rotate through 360° about its fulcrum. Fig. 2.7 shows a few examples, but there are many more including the windmill, the turbine, the bicycle's rear wheel, the brace and bit, etc. The basic parts are shown in Fig. 2.7(*a*); the wheel has a radius e and the axle l. The axle is fixed to the wheel and both rotate about an axis 0. Now if you look closely at this figure you will see that it is in fact a lever of the first kind, for the effort is exerting a clockwise torque Ee about the fulcrum 0, and the load is exerting an anticlockwise torque Ll about 0. So for equilibrium $Ee = Ll$, giving a mechanical advantage $L/E = e/l$. This is the answer we obtained earlier

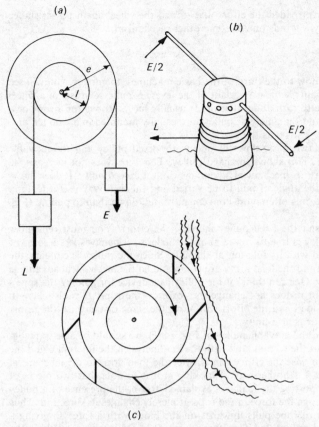

Fig. 2.7. The wheel and axle: (*a*) schematic, (*b*) capstan, (*c*) waterwheel.

for the simple lever. We should not be surprised therefore if the transmission factor for the wheel and axle were also the same as that for the simple lever of the first kind, i.e. $TF = L/e$. You can verify this from Fig. 2.7(*a*). For one complete clockwise turn of the wheel the rope sustaining the effort will unwind by $2\pi e$ and thus the effort will move through a distance $2\pi e$. At the same time the rope supporting the load will wind up by $2\pi l$ and thus the load will be raised through a distance $2\pi l$. The transmission factor is thus

$$TF = \frac{\text{Distance moved by load}}{\text{Distance moved by effort}} = \frac{2\pi l}{2\pi e} = \frac{l}{e}$$

One of Man's first uses for the elementary machine was the capstan (Fig. 2.7(*b*)), in which hand-spokes made up the wheel and the load was attached to a rope wrapped around the axle. Most of the old ships used this device for raising the anchor. In the waterwheel (Fig. 2.7(*c*)) the weight of the water provided the effort that turned the wheel against a resistance generated by a grindstone or some other mechanism.

The Pulley

Moving on now to the third of the basic machines, the pulley, you will see that once again we have a variant of the lever. A pulley consists of a wheel which is free to turn about its axle. It usually has a groove for a rope or a wire cable, and the pulley or pulleys are usually mounted in a wooden or a metal frame, known as the block.

There are two basic types of pulley: the fixed pulley and the movable pulley. Fig. 2.8(*a*) shows the fixed pulley. This type does not increase the applied effort; its mechanical advantage cannot exceed unity. It does, however, allow the angle of pull to be varied and this is a very useful feature. For example, it is often more convenient to pull down than to pull up (Fig. 2.8(*b*)).

Why doesn't the fixed pulley increase the effort? You must remember that the pulley is free to rotate about its axle, so it behaves like a lever of the first kind with its fulcrum at the axle. Since the pulley is circular, the effort arm *e* and the load arm *l* are equal and so the mechanical advantage will be unity. One can think of the pulley as a device for guiding the rope's direction. But it does not change the force on the rope. It is also easy to see that the load and the effort will move the same distance, so the transmission factor is also unity.

Turning to the movable pulley of Fig. 2.8(*c*), we should be able to recognise a lever of the second kind with the load between the fulcrum 0 and the effort. In this case the effort arm is twice the load arm, so the mechanical advantage is 2. Another way of looking at it is to note that two pieces of rope are supporting the load. As explained above, the presence of a pulley does not change the force in the rope, it merely changes its direction. Thus each section of rope pulls upwards on the block with a force *E*, giving a total upward pull of 2*E*. The effort then only needs to be one half of the load, so the mechanical advantage is 2.

If the mechanical advantage is 2 we would expect the transmission factor to be 0.5, and you can see this must be so if you consider what happens if the load is raised by 1 m. Both sections of the rope must shorten by 1 m, so the total movement of the rope will be 2 m. You should check this out for the crane in Fig. 2.8(*d*), which combines both a fixed and a movable pulley.

Some more complicated arrangements of pulleys are shown in Fig. 2.9. For simplicity the pulleys have been shown on separate axles, but in practice the fixed pulleys would run independently on one common axle and the moving pulleys on another, as in Fig. 2.9(*c*). Referring to Fig. 2.9(*a*), where there are two fixed pulleys and one movable one, it can be seen that the load is supported by three rope segments. Thus, since the effort *E* is transmitted through the full length of the rope (neglecting friction effects), then the effort only needs to be one third of the load. In general, the mechanical advantage will be equal to the number of segments of rope supporting the movable block. Check this out on Fig. 2.9(*c*). Similarly, the transmission factor will be the inverse of this number. For every metre that

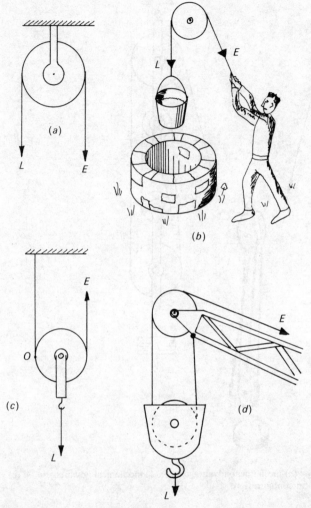

Fig. 2.8. Pulleys: (*a*) and (*b*) fixed and equivalent to levers of the first kind; (*c*) and (*d*) movable and equivalent to levers of the second kind.

the load is raised, each supporting segment of rope will have to shorten by 1 metre, and the effort will then move through a distance of *x* metres, where *x* is the number of supporting segments.

The Inclined Plane

We now turn to an entirely different type of machine, the inclined plane or wedge. Imagine that you have to get a garden roller to the top of a hill (Fig. 2.10(*a*)). Would you prefer to push it up a short steep road or up a

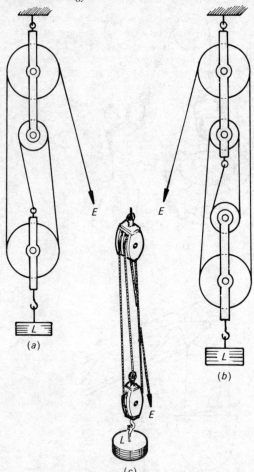

Fig. 2.9. Pulleys: (a) mechanical advantage = 3; (b) mechanical advantage = 4; (c) practical configuration.

longer gentle slope? No matter which route you choose you would still have to do the same amount of work (if friction is neglected). The weight has to be raised a certain height and the work required is the product of the weight and the height. The inclined plane helps us by acting as a machine, allowing us to match our work output to that required by the load (see Fig. 2.6).

We know from experience that it is less tiring to walk up a gradual slope than a steep one. Why is this? We can explain it by considering the energy requirements. If the roller weighs 500 N and it is raised vertically through 10 m, it would required $500 \times 10 = 5000$ J of work. If instead you were to push it up a slope 20 m in length then, since work is the product of force and distance, you would expect to exert a force of only $5000/20 = 250$ N.

The mechanical advantage of this peculiar machine would then be 500/250 = 2. Its transmission factor, the ratio of the load's movement to the effort's movement, would be 10/20 = 0.5. In general, neglecting friction, the mechanical advantage of an inclined plane is the ratio of the length of the slope to the vertical height (the cosecant of the angle of the slope). Thus the gentler the slope the greater the mechanical advantage.

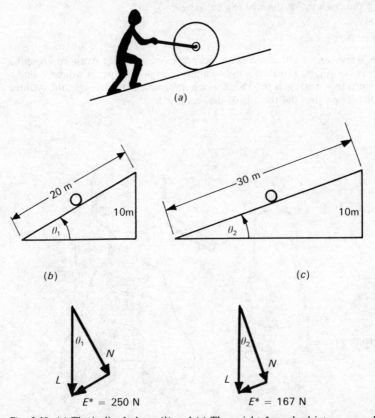

Fig. 2.10. (a) The inclined plane. (b) and (c) The weight L resolved into a normal component N and a parallel component E* for different slope lengths.

This can also be explained by considering the various forces acting on the roller. In Fig. 2.10(b) and (c) the weight L of the roller is shown broken into two components, one normal to the plane (N) and the other along the plane (E*). The normal component is entirely borne by the plane and does not affect the person pushing the roller (if friction is neglected). However, the pusher has to contend with E*, acting down the plane, and his effort E must be equal and opposite to E* if the roller is to move at constant velocity. The trigonometry confirms that the mechanical advantage $L/E = \operatorname{cosec} \theta$.

The inclined plane is called a passive machine because it does not move. But its brother, the wedge, is more active and finds use in most devices for cutting and piercing—axes, chisels, ploughs, nails. Its principle of operation is much the same as that of the inclined plane. The sloping face transforms a small force with large displacement at the end of the wedge to a large force with small displacement at the point of contact with the workpiece. In ideal circumstances the wedge will amplify the applied force by one half of the cosecant of the wedge's half-angle.

The Screw

A screw can be visualised as an inclined plane or wedge wrapped around a cylinder (Fig. 2.11(*a*)). If a right-angled triangle of paper is wound round a pencil then a screw is produced. The hypotenuse of the triangle, the inclined plane, becomes the threads of the screw.

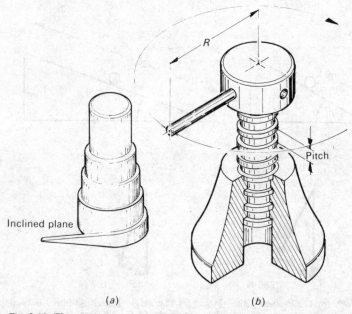

(*a*)   (*b*)

Fig. 2.11. The screw.

The pitch of a screw is the axial distance between the two adjacent threads. In Fig. 2.11(*b*) you can see that if the lever arm makes a full turn, the screw is raised through one pitch. We can then calculate the transmission factor TF

$$\text{TF} = \frac{\text{Load movement}}{\text{Effort movement}} = \frac{\text{Pitch}}{2\pi R}$$

where R is the lever arm.

If friction is neglected, the mechanical advantage MA is

$$MA = 1/TF = 2\pi R/\text{Pitch}$$

Consider, for example, using a jack to raise one corner of a car. The load is about 2500 N. Using a tommy bar 0.2 m long and a screw with a pitch of 2 mm, gives a mechanical advantage of $2\pi \times 0.2/0.002 = 628$. Thus we would only have to apply an effort of $2500/628 \cong 4$ N. But in order to raise the corner of the car through 0.5 m we would have to turn the jack $0.5/0.002 = 250$ times.

This example shows that the screw is not really a 'simple' machine, for its operation relies on the use of another machine, the lever. The example also shows that one basic feature of the screw is its ability to convert rotary into linear motion. Examples of this are numerous: the brace and bit, the corkscrew, the adjustable spanner, drills and so on.

Screws are also popular for fastening things together, and here friction is put to good use. Friction is a nuisance when we have to work against it; if we had taken it into account in the above example, the required effort would have been about 50 per cent greater. However, if there were no friction, the jack would unwind as soon as the handle was released—this is equivalent to a man sliding down a slippery inclined plane. So friction is important from this point of view, and without it a nut and bolt would be impractical as a fastening device, and screw jacks would have to be treated with caution.

2.3 Engines

The five basic machines we have just described were intended to supplement Man's musculature; they matched his capabilities to the job in hand. But there are many other forces in nature that could assist Man to do useful work.

Naturally-occurring forms of energy can be used to do work, and Man soon discovered how to tap these resources. He developed **engines**— machines for converting a naturally-occurring form of energy into a more directly useful form of work. For example, the windmill converted the naturally-occurring kinetic energy of the air into the kinetic energy of the grindstone. You will find that engines are often referred to as **prime movers** because they provide the energy that initiates the action of machines. The internal combustion engine converts the naturally-occurring chemical energy of oil into the kinetic energy of rotation of a flywheel. But this can be used as the prime mover for many difficult types of machines, such as motorcars, textile machines and agricultural machines.

Engines fall into two categories: those that use the pressures of hot gases to drive a piston along a cylinder, and those that use the energy of moving fluids to drive rotors. Let us examine the piston engine first.

2.4 Piston Engines

We have already traced the development of the steam engine in Chapter 1 so there is no need to repeat that here. It would be useful, however, to study the basic mechanism whereby the linear, up and down motion of a piston is converted to a rotary motion at the output of the engine. Fig. 2.12 shows the basic elements of a reciprocating engine. As the piston moves along the cylinder, the connecting rod drives the crank in a rotary manner. Note that when the output, the crank, rotates through one revolution the piston makes two strokes, a **stroke** being the distance covered by the piston as it travels from one end of the cylinder to the other.

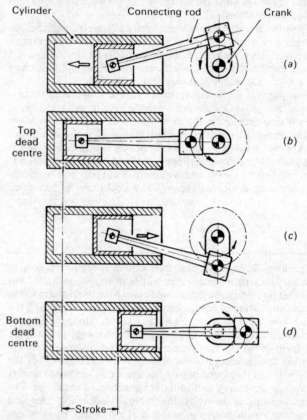

Fig. 2.12. From rotary to reciprocating motion.

The hot gases that drive the piston along the cylinder are generated by burning fuel and this can be done outside or inside the cylinder. If outside we have an **external combustion engine** such as the steam engine where the fuel (coal or wood) is burned in a boiler to produce the steam that provides the motive power. Today the majority of piston engines are of the **internal combustion** variety where the fuel is burned inside the cylinder. The petrol engine, the most commonly found of this type, uses an electrical spark to burn an air-petrol mixture in the cylinder. Internal combustion engines are popular because of their compactness; they do not need bulky boilers.

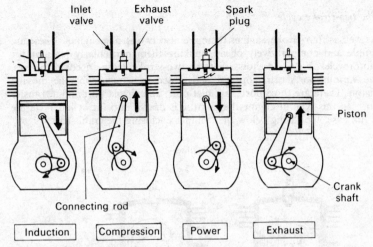

Fig. 2.13. The four-stroke cycle.

The burning fuel can be used to drive the piston from one end of its stroke to the other, and an examination of Fig. 2.13 shows that this will turn the crank through one half revolution. But how do we get it the whole way round? How do we keep the engine running? In order to understand this we must look at the engine cycle.

The four-stroke cycle

Most petrol engines in use today operate on a four-stroke cycle; that is, the piston makes four strokes during a complete cycle. The four strokes consist of (1) a suction or induction stroke, (2) a compression stroke, (3) an expansion or working stroke, and (4) an exhaust stroke. These are illustrated in Fig. 2.13. During the **induction** stroke an inlet valve opens and the piston, moving down, draws in a mixture of air and petrol from the carburettor. When the piston reaches the bottom of its stroke, the rotating crank starts to push it upwards, compressing the mixture. During this **compression** stroke both inlet and exhaust valves are tightly closed and at the top of the stroke the mixture is compressed by as much as 14 times. Everything is now set for the working stroke. The sparking plug ignites the fuel which, burning rapidly, produces hot gases that push the piston down the cylinder. Useful work has been done and it is now necessary to clear the cylinder of the spent fuel. The **exhaust** stroke commences, the exhaust valve opens and the upward moving piston pushes the burnt gases out into the exhaust system.

You will note that there is only one working stroke, and you may wonder where the energy comes from to drive the piston through its other three strokes. This is achieved by attaching a heavy flywheel to the output shaft. The energy imparted to the flywheel during each working stroke keeps it spinning and allows it to drive the piston during the other three strokes. The flywheel is really an energy-storing device.

The two-stroke cycle

As well as four-stroke engines there are also two-stroke engines. These are simple and comparatively cheap, making them particularly popular for motorcycles. Fig. 2.14 shows that the two-stroke engine needs no valves, the various ports being opened and closed in sequence by the moving piston. There are three ports: an inlet port, an exhaust port and a transfer port. In this type of engine the mixture is drawn into the cylinder via the crank case, the casing below the cylinder enclosing the crankshaft.

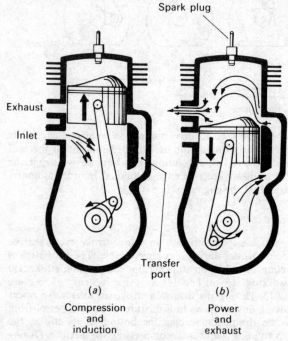

Fig. 2.14. The two-stroke cycle.

As the piston moves up the cylinder, compressing the fuel mixture, it simultaneously develops a suction in the crank case and this causes fresh fuel to be drawn in through the inlet port. At the top of the stroke, compression is maximum, the sparking plug fires, combustion takes place and the piston is driven downwards. As it descends it uncovers the exhaust port, through which the burnt gases are expelled, and the transfer port through which the fresh charge of fuel is transferred from the crank case. This movement of fuel takes place because the descending piston has pressurised the mixture in the crank case. You will note from the illustration that the inrushing transferred fuel helps to clear the cylinder of the remaining exhaust fumes. The engine has now finished one complete cycle consisting of two strokes—an upstroke and a downstroke.

The Diesel Engine

The sparking plug is a common device for igniting the air/fuel mixture, but another way of doing this was invented by Rudolf Diesel in 1892. Gases get hot when compressed, as a lot of us have experienced when using a bicycle pump. Indeed, this phenomenon was used in the early 1800s in the fire piston, which was, in a way, a forerunner of the match. The mechanism consisted of a long cylinder, not unlike a bicycle pump, with a small piece of tinder at its bottom end. A piston fitted tightly within this cylinder, and when it was rapidly forced into the cylinder the compression of the gas generated sufficient heat to ignite the tinder. Diesel applied this principle to his engine. It works on a four-stroke basis, similar to that of Fig. 2.13, but in this case the diesel engine draws in air alone, and compresses it to raise its temperature to over 500°C. At the point of maximum compression an injector sprays in diesel oil which ignites spontaneously in the hot air and drives the piston on its power stroke. Diesel engines can be modified to run on almost any kind of inflammable fuel, even coal dust, but most of them run on diesel oil.

2.5   Rotary Engines

The Windmill

In Chapter 1 we saw that wind power, in the form of a sail, was one of Man's first allies in his quest for increased power. The early windmills could have at best generated about 25 kW, although their poor design resulted in an output nearer to 6 kW. A lot of effort was devoted to improving their efficiency, particularly by John Smeaton in the middle of the eighteenth century. But windmills were soon to be displaced by the newer technologies such as the steam engine. Today, however, there is a renewal of interest in the windmill, for fuels are running out, and we have to look more closely at renewable energy sources such as wind, wave and sun. Profiting from what they have learned about efficient propeller blades for aircraft, engineers have greatly improved the design of windmills.

What power can we expect from a windmill nowadays? Let us test our knowledge of kinetics. Wind is merely air in motion. Air has mass, hence wind has kinetic energy, and we argued earlier that energy could be converted to work. Imagine a windmill where the area swept out by the blades is A. If the wind velocity is V, a cylinder of air of length V and area A will pass through the windmill in 1 second. Calling the density of air ρ, the mass of this cylinder of air will be $\rho A V$, and its kinetic energy will be $0.5 \times (\rho A V) \times V^2 = 0.5\, \rho A V^3$. (You will recall that kinetic energy is one half the mass times the velocity squared.) This energy represents the amount of work that the cylinder of air can do in one second. But you will recall that power is the rate of doing work, or work per unit time, or work per second. Thus the maximum power that can be extracted from the wind is $0.5\, \rho A V^3$.

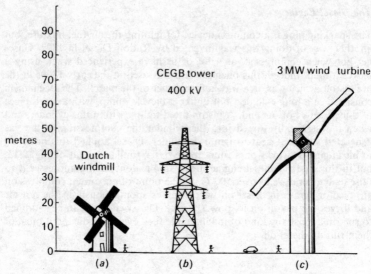

Fig. 2.15. Developments in windmills from (*a*) the Dutch windmill to (*c*) one considered for Orkney.

Taking	$\rho = 1.29$ kg/m^3
gives Power	$P = 0.000\ 013\ 7\ AV^3$

where P is kilowatts, A is square metres, and V is kilometres/hour.

In reality it has been found that the actual power available is only about 40 per cent of this theoretical maximum, so that a more realistic formula for power is

$$P = 0.000\ 005\ AV^3$$

For example, for a windspeed of 40 km/h, and a windmill of diameter 20 m, the power will be $0.000\ 005 \times \pi \times 10^2 \times 40^3 = 100.7$ kW.

This formula shows that if you double the diameter of the windmill's blades you should get four times the amount of power. The velocity of the wind is even more critical, for a doubling of wind speed will increase the output power by a factor of eight! So you may logically conclude that in order to get high power all we need to do is to build as large a windmill as possible and to place it in a position where the winds are very high. However, aerodynamic and economic considerations (including the cost of the structure) indicate that wind speeds greater than about 60 km/h should be avoided. Experimental machines generating over 1000 kW have been developed in France and the USA. The American machine, in Vermont, had a two-blade propellor 52.5 m in diameter and delivered a maximum of 1250 kW until one of its blades broke off during a storm. In the UK a 60 m diameter 3000 kW windmill has been considered for installation in Orkney. Fig. 2.15 gives an idea of its size.

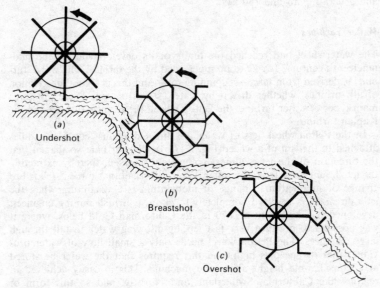

(a)
Undershot

(b)
Breastshot

(c)
Overshot

Fig. 2.16. Waterwheels.

The Waterwheel

The waterwheel was another important rotary prime mover to come to Man's assistance, and for some 1500 years it provided the main source of power for most industrial processes. We met the basic Vitruvian wheel in Chapter 1. Waterwheels were found in three forms: the undershot, the breast and the overshot (Fig. 2.16). The **undershot** wheel just dipped into the flowing stream and managed to capture only about 25 per cent of the available energy. In the **breast** wheel, water was delivered through a channel to a point about halfway up the wheel, and buckets on the paddles collected the water. The construction of these buckets required a good deal of engineering skill, for they had to fill quickly as the water entered them and they had to retain the water until they were at their lowest point of descent. The breast wheel was about twice as efficient as the undershot wheel. The final stage in the development of the waterwheel was the application of the water of the top of the wheel and this, the **overshot** wheel, could have an efficiency as high as 75 per cent. Each of the wheels was suitable for particular conditions; the undershot wheel when there was little fall; the breast wheel for falls of 2–5 m; and the overshot wheel where falls of 5–15 m were available.

In spite of their magnificent appearance these prime movers could not generate a lot of power. Even the great John Smeaton could only get about 35 kW from the machine built by him at Merthyr Tydfil in 1900. Attempts to obtain greater power outputs required the construction of very large wheels, such as the one at a Dublin papermill with a diameter of

26 m, and the one at Laxey in the Isle of Man, with a diameter of 22 m, and generating around 150 kW.

Water Turbines

The waterwheel had reached the limits of its development by the mid-nineteenth century. It was to be superseded by the **turbine**. The word 'turbine' is derived from *turbo*—a spinning top—and this is a good description of all turbines, whether driven by water, steam or gas. There are three main types of water turbine: the Pelton wheel, the Francis turbine and the Kaplan turbine.

In the **Pelton wheel**, jets of water are directed against cup-like paddles attached to the rim of a wheel (Fig. 2.17(*a*)). The cups are so shaped that the direction of the jet is almost completely reversed, thereby extracting the maximum force from the water. (You will recall that force is related to change of momentum. Change of momentum is a maximum when the jet's direction is reversed completely.) Pelton, a British mining engineer, developed his invention in 1870 in the Californian Gold fields, where it was possible to produce very fast jets by allowing water to fall through large distances. The Pelton wheel needs only a small flow of water providing that its speed is high, and this requires that the water be stored at a considerable height above the machine. This is easily achieved in regions like California, Switzerland and Norway, and so this form of turbine is quite popular in those regions. A substantial amount of power can be generated in the Pelton wheel. For example, an Alaskan wheel built in 1890 developed about 400 kW using water with a head of 140 m.

The **Francis turbine** extracts the power from the water in a different way. It consists of a rotor situated within a fixed outer ring of guide vanes (Fig. 2.17(*b*)). These vanes direct the water on to specially shaped runner blades on the rotor, and after the water has passed through these blades it escapes axially along the middle of the wheel. Francis developed his turbine in the 1840s. He depended a lot on the work of Fourneyron, who had invented a similar turbine in 1826 in which the water flowed outwards from a fixed inner wheel of guide vanes to a moving outer wheel. The Francis turbine has a large power capacity, 5000 kW being not uncommon. A major area of application has been in hydroelectric power stations.

In 1910 Kaplan invented yet another water turbine. He had been trying to improve the Francis turbine, which is not good with low heads of water, when he thought of the idea of using a ship's propeller in reverse. Instead of using a propeller to push water, why not use moving water to drive a propeller around? The **Kaplan turbine** (Fig. 2.17(*c*)) has four blades whose angle or pitch can be varied while the turbine is running, and this allows the engineer to extract the maximum amount of power as the water level varies. It is possible by this means to generate large amounts of power from large volumes of water at low head. In the Soviet Union, eight sets, each of 28 MW capacity, were installed on the River Svir in 1929, and more recently the French Rance tidal scheme used 24 sets, each of 10 MW capacity.

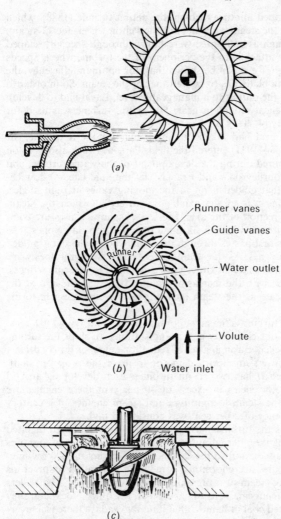

Runner vanes
Guide vanes
Water outlet
Volute
Water inlet

Fig. 2.17. Water turbines: (*a*) the Pelton wheel, (*b*) the Francis turbine, (*c*) the Kaplan turbine.

Steam Turbines

Carl Gustave Patrick de Laval (1845–1913) was the first man to produce a practical steam turbine (Fig. 2.18(*a*)). It was a simple turbine with an S-shaped rotor, working on the same principle as Hero's turbine of 2000 years earlier. Laval's S turbine ran at very high speeds, up to 40 000 rev/min, and this presented difficult balancing problems. However, in 1883 he considered that all the problems had been mastered and he took out a patent to use his turbine for driving milk separators.

Laval also developed another turbine, the action turbine (1888), which to some extent was the steam equivalent of the Pelton wheel. Jets of steam, with velocities as high as 1500 m/s, were blown through specially shaped nozzles on to vanes attached to the periphery of the turbine wheel. Speeds again were very high, for it can be shown that, for optimum efficiency, the speed of the rotating blade has to be half that of the steam. So in order to reduce the speed at the output to a manageable level, Laval had to develop a special double helical reduction gear. His action turbine was used for generating electricity (4–400 kW).

The year after Laval had taken out his patent for the S turbine, Sir Charles Parsons (1854–1931) entered the field. Parsons' work also built on the knowledge acquired during the development of the water turbine, and this turbine, like Fourneyron's and Francis', let the fluid (steam) flow between fixed vanes that guided it on to the moving vanes at right angles. Parsons' first turbine was of the axial flow variety in which the steam moved along the direction of the axle. Fig. 2.18(*b*) shows a pressure-compounded turbine with ten wheels—six small ones and four large ones. The high-pressure steam first passes through the small wheels where it is guided by six sets of guide vanes. As the steam traverses this section it does work and loses pressure, and as we all know a gas expands as its pressure reduces. So, in order to cope with the increasing volume of steam, the size of the turbine has to increase as the steam moves from the high-pressure to the low-pressure end.

Today the steam turbine is the most popular means of driving the largest ships and many of our electricity-generating stations. In ships the turbine may drive the propellers through a set of reduction gears, or it may drive a generator which powers an electrical motor to drive the propeller shaft. The *Queen Elizabeth II* has two steam turbines, each developing around 40 000 kW. With efficiencies in excess of 40 per cent these engines are vastly superior to the steam locomotives and steam engines of a century ago, when an efficiency of 2 per cent was considered high.

In power stations the turbines have to run at a constant speed dictated by the frequency of the alternating current. In most European countries this is 50 Hz (cycles/sec), and so the turbines have to run at 3000 rev/min. Power station turbines are enormous, sometimes generating as much as 600 MW, and using steam at a pressure of 160 atmospheres. The boilers that generate this steam can be as big as a 30-storey office block, and they burn oil or pulverised coal in flames 30 m long. Nowadays there is a growing move towards the nuclear power stations in which the heat generated by nuclear fission or fusion is used to boil the water.

Gas Turbines

The gas turbine, like the steam turbine, has one main moving part, the rotor. But in this case the rotor does two different jobs at once: it provides power to drive an external load, such as the generator, whilst at the same time driving its own **compressor** (Fig. 2.19(*a*)). The compressor rotor turns inside a closely fitting casing which has sets of stationary blades so placed that they just fit between the revolving rotor blades. When the rotor is spinning at high speed, air is drawn in and compressed as it passes through

(a)

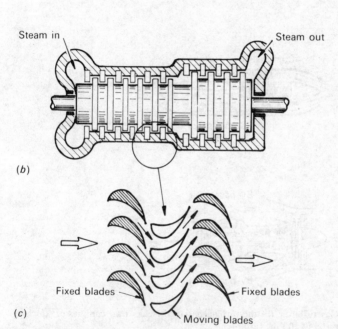

Steam in

Steam out

(b)

(c)

Fixed blades

Fixed blades

Moving blades

Fig. 2.18. Steam turbines: (a) Laval's turbine, (b) Parsons' compound turbine, (c) fixed and moving blades.

the sets of moving and fixed blades. In modern engines it is possible to increase the pressure of the air by a factor of 30 by using 12 sets of blades.

When the air is compressed it enters a combustion chamber where it is burned with paraffin or diesel oil. (Can you see the analogy with the reciprocating engine?) The combustion process generates very high temperatures in the engine, so it is necessary to use special heat-resistant or refractory material for the construction of the combustion chamber and for the turbine blades.

These hot gases then flow through the turbine section which, like the compressor, consists of several sets of fixed and moving blades. As in the

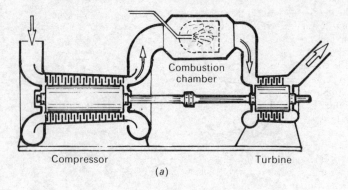

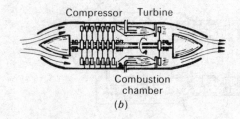

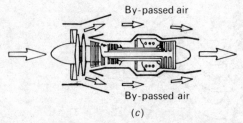

Fig. 2.19. Gas turbines: (*a*) the turbine has to drive its own compressor; (*b*) the turbo-jet; (*c*) the turbo-fan.

water and steam turbines, the function of the fixed guide vanes is to direct the streams of hot gas on to the rotor blades at an angle of 90 degrees.

The combustion process results in two main forms of useful energy: the energy in the rotating shaft, and the energy in the jet of high-velocity gas that is ejected from the rear of the engine. In a turbo-prop engine the shaft power predominates and is used to drive, through a reduction gear, the propeller of the aircraft. For example, many light transport aircraft use the Pratt and Whitney PT6 engine in which the turbine runs at 30,000 rev/min and the propeller at 1320 rev/min. The turbo-shaft engine also uses the shaft power of the gas turbine, and is used to drive ships, hydrofoils, hovercraft, generators and gas-pumping units.

In the turbo-jet engine (Fig. 2.19(*b*)) the turbine section extracts only a part of the energy of the gas flow. Thus there is a lot of energy left in the efflux from the engine, and by action and reaction, since the jet goes backwards, the engine is pushed forwards. Whittle and von Ohain developed the first jet engines independently in the 1930s. Whittle's engine first ran in 1937 and it was used in the first British jet plane, the Gloster E28/39, in 1941. Two years earlier, the world's first jet plane, the Heinkel He 178, had flown for the first time at Warnemunde.

The modern jet engine is impressive. Concorde is propelled by four Rolls Royce/SNE CMA Olympus engines which together generate about 600 000 N of thrust at take-off. But jet engines can be very noisy and a lot of work has been carried out to try to fill the gap between noisy high-speed turbo-jet engines and the quieter low-speed turbo-prop. The problem arises because the turbo-jet throws back a very fast jet of hot gas that tends to tear the air apart. The thrust of this type of engine is related to the product of the mass of gas expelled and its velocity, so if the velocity is reduced to eliminate the noise problem, the mass of gas has to be increased in order to maintain the thrust. This is achieved in the turbo-fan engine, shown in Fig. 2.19(c), which gets its name from the large fan mounted at the front of the engine. The fan is basically an oversized compressor, and to some extent acts like a conventional propeller. A large proportion, as high as two thirds, of the incoming air by-passes the core of the engine, but is reunited with the high-velocity jet at the rear of the engine. The combined effect is to produce a greater mass flow at a smaller velocity, and hence with less noise.

2.6   Linear Engines

We have now completed our survey of reciprocating and rotary engines. One type of engine remains—the **rocket** engine. Its operation relies on action and reaction, formulated by Newton in his third law of motion—every action produces an equal and opposite reaction.

Think of a cannon gun; as the cannon ball is projected the cannon kicks in the other direction. Similarly in the rocket engine; as the stream of hot gas is ejected the rocket is 'kicked' in the other direction. There is another easy way of envisaging this phenomenon. Imagine a closed cylinder (Fig. 2.20(a)) filled with high-pressure gas. We know from basic physics (Pascal's Law) that the pressure acts equally on all the inner walls, so the forces acting are balanced. But imagine that one end of the cylinder is removed. The gas will flow out rapidly, its pressure on the opposite end will no longer be balanced, and the cylinder will begin to move. This is the basis of a rocket's action, and it illustrates an important fact—a rocket needs no surface or fluid to bear against, so it can operate in the vacuum of outer space. Indeed, the rocket engine is the only existing propulsive system that can function beyond the Earth's atmosphere. (Several new schemes are under consideration including the photon rocket which will provide thrust from a beam of concentrated light.)

How are the gases produced? They are generated by the combustion of a fuel/oxygen mixture, sometimes in solid form, sometimes in liquid form. In solid propellant rockets the fuel/oxygen mixture takes the form of a hollow

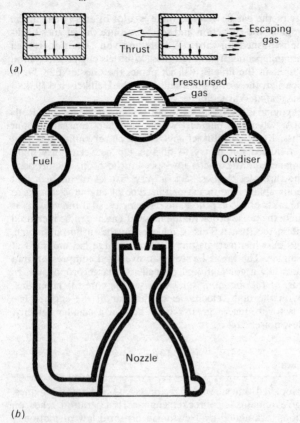

(a)

(b)

Fig. 2.20. Rocket motors: (a) the pressure imbalance causes the thrust; (b) the generation of the hot gases.

cylinder perforated with holes. A typical solid propellant consists of nitrocellulose plasticised with nitroglycerine and mixed with certain stabilising products that slow down the rate of burning. In liquid propellant systems (Fig. 2.20(b)) the most common oxidiser is liquid oxygen. As far back as 1926, Robert Goddard used petrol as the fuel. Later the V2 rocket used alcohol but nowadays liquid hydrogen is becoming extremely popular.

We can calculate the thrust of a rocket if we know the mass of fuel M burnt per second and the velocity V of the exhaust gases. Newton's second law states that force F is equal to rate of change of momentum. In 1 second the momentum of the gas changes by MV; therefore the force acting upon it is MV. By action and reaction this force MV also acts on the rocket, but in the opposite direction. For example, if 50 kg of propellant are burnt each second, and the exhaust velocity is 3000 m/s, then the thrust is 150 000 N.

This shows how important it is to keep the exhaust velocity as high as possible, and liquid propellants are the best in this respect. The liquid hydrogen/liquid oxygen mix gives velocities of around 3800 m/s whilst the best solid propellants only manage a mere 2400 m/s. Nevertheless, the fact that a solid propellant is always 'ready to go' has made it popular for missiles. For example, the Minuteman can be stored fully charged and can be launched in less than 1 minute.

The latest space venture uses a combination of solid and liquid propellants. The Space Shuttle flight system uses two solid rocket boosters, each with a thrust of 11.8 million Newtons. Each of the Orbiter's three main liquid rocket engines develops a thrust of 2.1 million Newtons, using a liquid oxygen/liquid hydrogen mixture.

2.7 Transmission of Power

In some engines, such as the aircraft gas turbine, the power is generated at the place where it is needed. In many other cases, this is not so. For example, the power developed in the car engine has to be transmitted to the wheels. An even greater problem is the transmission of the power of the steam turbine from the power station to the home or the factory. We shall look briefly at transmission of power by mechanical, by fluid and by electrical means.

Mechanical transmission

Belt drives were used to transmit power around the early factories. Nowadays they can be found in vacuum cleaners, motor mowers and lathes, but perhaps their best-known application is found under the bonnet of the car where a belt is used to drive the fan and the alternator from the crank shaft. As well as transmitting power from A to B, the belt drive can also be used to transform the power (see earlier). Fig. 2.21(*a*) shows a belt connecting two wheels A and B. Wheel A of radius r is the input and wheel B of radius R is the output. One clockwise revolution of A will move the belt through a distance of $2\pi r$. Since it takes a belt movement of $2\pi R$ to turn wheel B through one revolution, then a movement of $2\pi r$ will turn wheel B through $(2\pi r)/(2\pi R) = r/R$ revolutions. The transmission factor is thus r/R, and the output will rotate at r/R times the speed of the input. Now power is the product of torque and angular velocity,* so, if there is no power loss in the transmission, the output torque will be R/r times the input torque. For example, if the diameter of wheel B is twice that of A, then the output speed will be one half of the input and the output torque will be twice that of the input.

In Fig. 2.21(*a*) the output and the input wheels rotate in the same direction. A reversal can be achieved by using a crossed belt, as in Fig. 2.21(*b*). Since belts rely on friction for their effective operation, there is a limit to the force that they can transmit. This aspect of performance can be improved by using a toothed belt but often this limitation requires the use of a more positive drive such as a chain and sprocket or a gear train.

In a **chain and sprocket drive**, such as that on a bicycle, the links of the chain are all of the same size, so the sprocket teeth have to be the same size on both drive and driven wheels. The transmission factor is found by

---

* This is the rotary equivalent of the formula that was developed on page 32.

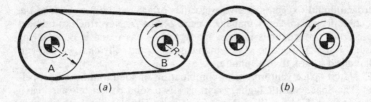

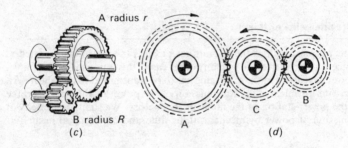

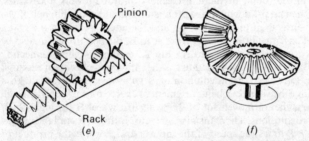

Fig. 2.21. Mechanical transmissions: (*a*) belts; (*b*) crossed belts; (*c*) spur gears; (*d*) gear train with idler C; (*e*) rack and pinion; (*f*) bevel gears.

dividing the number of teeth on the input sprocket by the number on the output sprocket.

The **gear train** is the most commonly used form of mechanical power transmission. Fig. 2.21(*c*) shows two spur wheels in mesh. We showed earlier that for belts the transmission factor between the input wheel A and the output wheel B is r/R. The same applies for gears, but since the teeth on both wheels are the same size, the ratio of the radii is also equal to the ratio of the number of teeth on the wheels. Thus the transmission factor is equal to the number of gear teeth T_a on A divided by the number of teeth T_b on B, and the wheel B will rotate at T_a/T_b times the speed of the wheel A. Again, assuming no energy loss, the torque at the output shaft will be T_b/T_a times that at the input.

Note that wheel B rotates in the opposite direction to wheel A. If output and input have to turn in the same direction, a third wheel C can be used (Fig. 2.21(d)). We can write

$$\text{Speed of B/Speed of C} = T_c/T_b$$
and $\quad \text{Speed of C/Speed of A} = T_a/T_c$

$$\therefore \quad \frac{\text{Speed of B}}{\text{Speed of A}} = \frac{\text{Speed of B}}{\text{Speed of C}} \cdot \frac{\text{Speed of C}}{\text{Speed of A}} = \frac{T_c}{T_b} \cdot \frac{T_a}{T_c} = \frac{T_a}{T_b}$$

It is interesting to note that the wheel C does not affect the speed of the output; it simply changes its direction. For this reason it is called an idler wheel.

Gears are excellent examples of power-transforming devices, and we are all aware how they are used to good effect in gearboxes. The car engine is at its best when running at high speed, but we cannot always let the car itself run at high speed. So a gearbox is used to transform the power developed by the fast-running engine to a form more suitable for use at the wheels, where demands can vary from a standing start to a fast cruise.

There are many different types of gears. Fig. 2.21(e) shows the **rack and pinion** drive that is used to convert rotary to linear motion, or vice versa. This is used, for example, in some car steering systems where the pinion is connected to the steering wheel and the rack is used to turn the car wheels. Fig. 2.21(f) shows a pair of **bevel** gears which can be used when it is necessary to turn power around a corner. An example of their use is the transmission of power from the drive shaft in a car to the back axle.

Hydrostatic Transmissions

Pressurised fluids also provide a very flexible form of power transmission. The development of such systems owes a lot to the Frenchman Pascal, who proved in 1648 that the pressure in a fluid is transmitted equally in all directions. More than a century later, Bramah used the principle to demonstrate that pressurised fluids can be used to amplify force and transform power. Imagine two cylinders of areas A and a, connected by a pipe (Fig. 2.22(a)). Each cylinder has a close-fitting piston that is free to move up and down. What force f is required on the smaller piston to hold a load F resting on the larger piston?

We can work it out this way. Assume that the system is in equilibrium. Thus the force f develops a pressure f/a on the face of the smaller piston. But Pascal's Law states that the pressure in a fluid is transmitted equally in all directions. Therefore a pressure $p = f/a$ acts on the walls of both cylinders, on the walls of the pipe, and on the larger piston of area A. Now if pressure p acts on a piston of area A, that piston will generate a thrust $F = pA = fA/a$. Thus the thrust of the larger output piston exceeds that of the input by a factor A/a. You can see the tremendous possibilities; if A is 100 times a, then a push of only 100 N would be sufficient to raise a car weighing 10 000 N.

This is too good to believe! There is a catch of course. Once again this machine is limited by the First Law of Thermodynamics. We cannot get

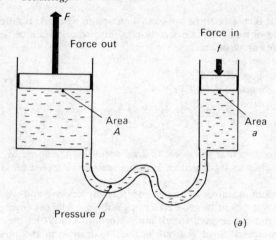

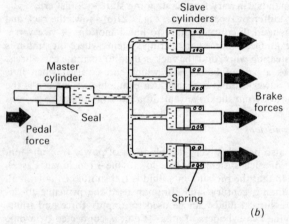

Fig. 2.22. Hydrostatic transmission: (*a*) Bramah's principle; (*b*) car brake system.

more work out than we put in, and since work is the product of force and distance, we would expect the movement of the output piston to be reduced by the same factor by which its force was increased. We can see that this is so by imagining that the small input piston is pushed down through 10 cm. This will cause an amount of fluid of volume $0.1a$ cubic metres to pass along the connecting pipe to the larger cylinder, where it will displace the output piston by $(0.1a)/A$ metres. The output piston's movement is then a/A times that of the input, so in the above example, where $A = 100a$, the output motion will be one hundredth that of the input. (This system is sometimes called the hydraulic lever. Can you see why? Bramah might have said: 'Give me an output piston large enough and I will move the Earth!')

Besides the ability to produce large forces, another advantage of a fluid transmission is the fact that the power can be passed along flexible pipes—so it can be turned around corners quite easily. For example, the car braking system was at one time wholly mechanical. The transmission linkages were complex in order to allow for the relative movements caused by steering and suspension, and to ensure that brake forces were balanced. The hydraulic system of Fig. 2.22(b) proved to be much better. It ensured that brake forces were equal and that forces were sufficiently high. The use of flexible pipes eliminated problems arising from relative motion of the various parts of the car.

Nowadays oil and air are the most popular forms of fluids for these systems. Oil is used in hydraulic systems and air in pneumatic systems. In some hydraulic systems, the small input piston in Fig. 2.21(a) takes the form of a hand pump such as that used for jacking up cars. But in most cases the input is supplied by a hydraulic pump or an air compressor. The output can be linear, driven directly by a piston and cylinder as shown in Fig. 2.22, or it may be a rotary hydraulic or pneumatic motor. Modern applications include the operation of aircraft control surfaces, excavating machinery, ship steering gear and forges for the metal-processing industries where forces up to 5 million Newtons (5 MN) are required. A frivolous application is illustrated by the latest American craze. You can have your old car, minus its engine, compressed into a block that can be used as a coffee table.

An important point to note is that hydraulic pumps and air compressors are not engines (or prime movers). They do not convert a naturally-occurring form of energy into a more useful form. They are energy-transforming devices that themselves require to be driven by an engine. They transform the input energy into fluid form which in turn can be reconverted to mechanical energy at the output, whether it be a piston and cylinder or a motor.

Electrical Transmission

Electricity **generators** fall into the same category; they convert mechanical energy into electrical energy—but they are not prime movers. In 1820 the Danish scientist Oersted discovered that a magnetic field was generated when an electric current flowed in a wire. Later, in 1831, Michael Faraday demonstrated that the reverse was also true: his simple generator showed that a current could be made to flow in a wire by turning a magnet around it. Faraday pictured a magnetic field between the poles of a magnet as consisting of 'lines of force', and he proved that when a coil or loop of wire was passed through these lines of force a voltage was generated whose magnitude was proportional to the rate at which the lines were cut by the coil.

The principle of the generator can be explained with the help of Fig. 2.23(a), which shows a coil of wire being rotated between the poles of a magnet. Two copper slip rings rotate with the coil, each end of the coil being connected to a slip ring. The generated voltage is taken off and applied to the external circuit by two spring-loaded graphite blocks (called brushes) that press against the slip rings.

The voltage generated by such a device is not constant. It alternates as

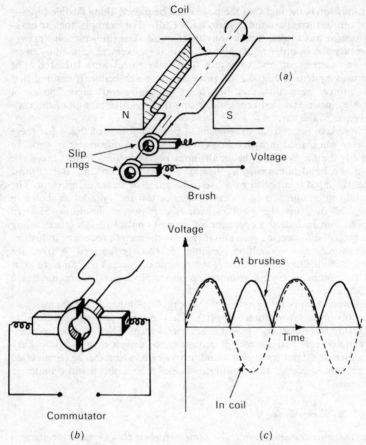

Fig. 2.23. Generation of electricity: (*a*) coil and slip rings rotating between magnets; (*b*) the commutator; (*c*) voltage waveforms at coil and at brushes when commutator is used.

the coil rotates, because the rate at which the magnet's lines of force are cut varies as the coil rotates. Fig. 2.24(*a*) shows that when the plane of the coil lies along the lines of force, a small rotation will cut quite a few lines of force, whilst for the plane of the coil at right angles to the lines of force (Fig. 2.24(*b*)) a similar rotation will cut a smaller number of lines. Thus the generated voltage for the former case will be larger than for the latter. A further 90 degrees of rotation (Fig. 2.24(*c*)) will bring the coil once again parallel to the lines of force, where the rate of cutting the lines will again be high. But the coil is now cutting the field in a different direction from the case of Fig. 2.24(*a*), so the generated voltage is of different sign. The resultant voltage variation is shown in Fig. 2.23(*c*). Because of the alternating nature of the voltage, this type of generator is called an **alternator**. The frequency of the alternating current (a.c.) is the number of complete cycles per second (Hertz or Hz)—50 Hz in European and 60 Hz in American power supplies. The alternators in power stations are designed to produce three phases: three of the above single phases, each reaching its maximum value in turn.

Lines of force

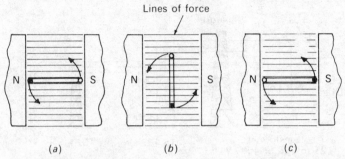

(a) (b) (c)

Fig. 2.24. A coil cutting lines of force. At (*a*) and (*c*) more lines are being cut in a given time than at (*b*).

A unidirectional or direct current (d.c.) can be achieved by using a commutator instead of slip rings to take the voltage off from the rotating coil. An elementary **commutator** consists of a split ring of copper, the two halves of which are insulated from each other and joined to the ends of the coil (Fig. 2.23(*b*)). The brushes are positioned so that the change-over of contact from one half of the commutator to the other occurs when the coil is at right angles to the lines of force. This is the position where the coil voltage changes sign, so the voltage at one brush is always positive and the other negative. The resultant current variation is shown in Fig. 2.23(*c*).

The action of the commutator relies on two negatives making a positive. Imagine two objects whose temperatures could continuously change from hot to cold, one always being hot while the other was cold. If we put our left hand on the hot object and our right hand on the cold but changed hands every time the bodies changed temperature, then we would always have our left hand on the hot object and our right on the cold one, and heat would always flow through our bodies from left to right. Our hand switching is an analogy of the action of the commutator.

In practice there are many coils placed at regular spaces around a moving rotor in a d.c. generator (often called a **dynamo**). This not only produces more electrical power, but in addition the slight phase differences between the voltage generated in each coil give a resultant total voltage that is nearly constant. This sort of dynamo is popular in motorcars where it is driven by the engine and supplies the electrical system and charges the battery.

We saw earlier how pneumatic or hydraulic systems are convenient ways of transmitting power. Electricity is even more convenient, and the evidence for this is there for all to see. Every house has an electricity supply to drive vacuum cleaners, to light lamps and electric fires. The distribution of this electricity throughout the country is a complex task; for example, the National Grid in Britain (Fig. 2.25) is a network of cables mostly supported on pylons, connecting over 200 power stations. In the largest power station a.c. is generated at 25 000 V (50 Hz) and this is stepped up to 275 000 V or 400 000 V for transmission over long distances. This is subsequently reduced in substations for distribution to local consumers—240 V for homes, 33 000 V for heavy industry, and 11 000 V for light industry.

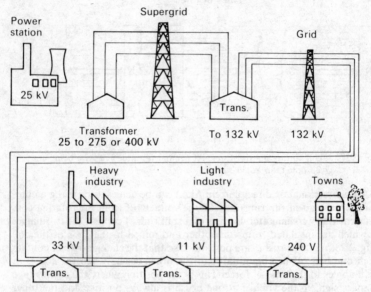

Fig. 2.25. The National Grid.

High voltages are used for transmission in order to reduce power losses. It can be shown that the power loss in a line is $(P/V)^2 \times R$, where P is the power delivered, V is the voltage and R is the resistance of the line. This power loss manifests itself as heat and is demonstrated, in the extreme case, in the electric fire. It can be reduced by making the resistance of the wire low, but this requires the diameter of the copper cables to be increased, and copper is expensive. So it is best to increase the voltage, especially since doubling it gives a fourfold decrease in power loss. But how is the voltage increased? It is not feasible to generate electricity at voltages higher than 25 000 V, but fortunately the **transformer**, invented by Faraday, allows alternating voltages to be stepped up or down with relative ease.

A transformer uses two coils (Fig. 2.26) called the primary and secondary windings, wound on a soft iron core. When a.c. is applied to the primary, an alternating magnetic flux is produced in the iron and this induces a voltage in the secondary. The voltage E_s in the secondary is approximately given by

$$\frac{E_s}{E_p} = \frac{\text{Volts in secondary}}{\text{Volts in primary}} = \frac{\text{Secondary turns}}{\text{Primary turns}}$$

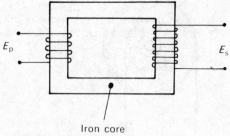

Iron core

Fig. 2.26. A transformer.

So a step-up transformer has more coils of wire on the secondary than on the primary. Note, however, that we can never escape the First Law of Thermodynamics. Electrical power is the product of voltage and current, so if the voltage goes up in the secondary, the current must go down by the same factor. If you consider voltage to be the analogue of force, and current to be the analogue of displacement, you should be able to see why the transformer is often referred to as an electrical lever.

We are concerned in this chapter with mechanisation, so let us conclude by briefly referring to electric motors. The principle of operation of the **d.c. motor** is simply that of a reversed dynamo. The following analogy may help the reader to understand how it works. Its operation relies simply on magnetic repulsion and attraction, and it can easily be demonstrated with the help of a compass needle and a bar magnet. Bring the South pole of the bar magnet towards the North pole of the needle. The needle will quickly swing round towards it, but when it comes nearly into line with the bar magnet, quickly turn the latter thus pointing its North pole towards the needle. This pole now repels the North pole of the needle, causing it to continue on its circular path. With a little bit of practice you can keep the needle spinning.

This is how a d.c. motor works, but in place of the compass needle there is a moving coil of wire, and in place of the bar magnet there is a large electromagnet. It is not easy to keep changing the poles of this large magnet, but the same effect can be achieved by changing the direction of current in the coil at each half revolution by means of the commutator.

The commonest form of **a.c. motor** is the induction motor. Its action depends on the fact that a moving magnetic field can set a neighbouring conductor into motion. In the large induction motor the moving magnetic field is achieved in the following way: three pairs of fixed electromagnets are spaced at equal angles round a conducting motor, and each pair is connected to one of the phases of a three-phase supply. This produces the effect of a rotating magnetic field which the rotor attempts to follow. An interesting feature of the induction motor is that it can be made in linear form; that is, with the stator windings arranged in a line rather than a circle. In this case the magnetic field moves along the line of the windings and any suitable piece of conducting material will be carried along with it. These linear motors are finding increasing use in new transport systems, and we will meet them again in Chapter 8.

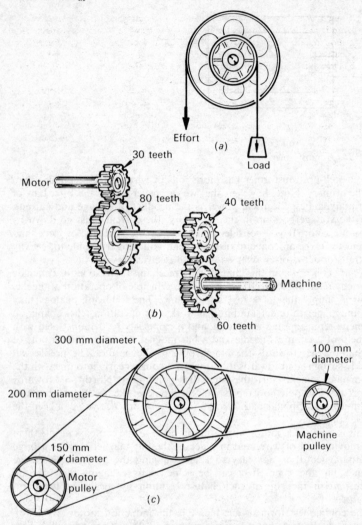

Fig. 2.27. Diagrams for exercises: (*a*) wheel and axle; (*b*) gear train; (*c*) pulley system.

2.8   Exercises

Numerical answers are given in Appendix 1.

1. (*a*) Explain the difference between energy, work and power. (*b*) An 80 kg man climbs the stairs to the top of a 50 m high building. What work has he done? (*c*) If the climb took 2.5 minutes, what power was required? (*d*) How does the man's energy differ at the finish of his climb? (*e*) If he slides down the bannister what happens to the work that was done during the climb?

2. (*a*) Describe the three basic levers, giving two examples of each. (*b*) Define mechanical advantage and transmission factor. (*c*) In the wheel and axle arrangement shown in Fig. 2.27(*a*) the wheel is 1 m diameter, the load is 100 N and the effort 20 N. Calculate the diameter of the axle if the efficiency is 100 per cent, the transmission factor, the mechanical advantage, the load if the efficiency fell to 80 per cent.

3. (*a*) List four differences between a spark ignition engine and a compression ignition engine. (*b*) Sketch the sequences of events that occur in these engines. (*c*) How does the two-stroke cycle differ from the four-stroke?

4. (*a*) Compare the historical developments of the windmill and the watermill, listing their relative advantages and disadvantages. (*b*) A windmill rotor has a diameter of 20 m. If the average windspeed is 50 km/h determine the available power.

5. (*a*) What is a prime mover? Give five examples. (*b*) What advantages are common to water turbines, steam turbines and gas turbines? Compare the power available from these turbines and give typical applications.

6. Give two examples of the transmission of power by (i) mechanical means, (ii) hydrostatic means, (iii) electrical means. Compare the advantages and disadvantages of these methods of power transmission.

7. Fig. 2.27(*b*) shows the drive for a machine tool. If the electric motor develops 5 kW at 800 rev/min, calculate (i) the torque at the motor shaft, (ii) the speed and the torque at the machine spindle. Assume 100 per cent efficiency. (*b*) If the same motor drives the system of pulleys and belts shown in Fig. 2.27(*c*), determine the speed and torque at the output if the efficiency is 80 per cent.

8. (*a*) An engine drives a shaft at 1000 rev/min. It is required to drive a parallel shaft in the same direction at 3000 rev/min. Sketch two suitable positive drives. (*b*) In an agricultural machine, belts are used to transmit power from shaft A to shafts B and C. A turns at 300 rev/min and B and C have to turn at 100 and 50 rev/min respectively. A pulley of 150 mm diameter is already fitted to shaft A, and one of 300 mm diameter to shaft C. Sketch a suitable system of pulleys and belts, and calculate the diameter of the extra pulleys that will be needed.

9. (*a*) Explain the operation of Bramah's hydraulic press. (*b*) A car hoist uses a hydraulic press to lift cars bodily for servicing. If the two pistons have diameters of 300 mm and 40 mm, calculate the force required from the operator to lift a 1000 kg car. (*c*) If the car is raised 2 m what work does the operator do (assuming 100 per cent efficiency)? (*d*) If the operator can at best generate a steady 0.2 kW, what is the shortest time in which he can raise the car? (*e*) What is the rate of flow of oil from the small cylinder to the large cylinder during this operation?

10. (*a*) Explain why it is necessary to use a high voltage when transmitting electrical energy. (*b*) In the British system what voltages are required for (i) light industry and (ii) domestic purposes? (*c*) Sketch a transformer for reducing voltage from level (i) to level (ii). (*d*) Describe how useful energy may be lost in a transformer.

2.9 Further Reading

Damon, A., Standt, H. W., and McFarland, R. A., *The Human Body in Equipment Design*, Harvard University Press, 1971.
Machines, Time Life Books, 1969.
Reay, D. A., *History of Manpowered Flight*, Pergamon Press, 1977.
Ross, D., *Energy from the Waves*, Pergamon Press, 1979.
Selected technology topics, Open University, PET 271 Block 3(9, 1S).
Strandh, S., *Machines: An Illustrated History*, Artists House, 1979.

'The prospects of the generation of electricity from wind energy in the UK', Department of Energy, *Energy Paper No. 21*, 1977.
Walker, C. R., *Technology, Industry and Man*, McGraw-Hill, 1968.
Whitt, F. R., and Wilson, D. G., *Bicycling Science*, MIT Press, 1977.

3
Technology Extends Man's Senses and Communication

3.1 Introduction

In this chapter we shall examine the ways in which technology has extended our **senses** and our means of **communication**. Sensing and communicating are related. Our senses determine the information that we extract from the real world; they are part of our communication link with reality. The technology of communication, on the other hand, allows us to pass information back to the real world.

3.2 The Six Senses

Without **sensors** we could not regulate our internal bodily functions, nor could we communicate with the outside world. Many different sense organs have evolved and, in general, each of them is capable of responding to only one form of stimulus. (You can imagine the confusion that would arise if our eyes responded to sound as well as to light.) Taste, smell, sight and hearing make up the so-called **special senses**, with touch completing the classical five. You will have heard of people being ascribed a 'sixth sense' when they display an ability to draw conclusions about the outside world without apparently using any of the above five senses. This ability is also referred to as extra sensory perception (ESP)—perception requiring additional senses—but I leave you to draw your own conclusions about the scientific credibility of ESP. It is interesting, however, to note that scientists do indeed find it convenient nowadays to designate another sense, a sixth sense—the sense that informs us of the movements and orientation of our own body. We shall hear more about this later.

Receptors

Every sense organ is a specialised structure consisting of one or more receptor cells and accessory tissues. In the eye, for example, the receptors are the rods and cones, and the accessory tissues are the lens, iris, cornea, etc. A sense organ has two tasks to fulfil: it has to detect and it has to transmit this information to the **central nervous system** (the brain and spinal cord).

There are four main types of receptor:

(1) **Mechanoreceptors** are sensitive to movement—stretch, twist, compression. They can be found in the skin, in the ear, in the muscles.

(2) **Photoreceptors** are sensitive to the visible part of the electromagnetic spectrum.

(3) **Chemoreceptors** react to particular combinations of chemicals and form the basis of our senses of taste and of smell.

(4) **Temperature receptors** detect heat and cold, and are commonly found in the skin.

These receptors detect changes in the environment, but before the body can respond to these changes information has to pass from the receptors to the central nervous system. The **neuron** does this job. The neuron is a nerve cell, the simplest element of nervous action (Fig. 3.1). It is a single cell with three important subdivisions: the **dendrites**, the **cell body** and the **axon**. Its shape is similar to that of an extremely tall tree, the branches or dendrites springing from a cell body that is connected to the top of the trunk-like axon. Electrochemical impulses from other cells are received by the dendrites, and the axon transmits these impulses to the dendrites of other cells or to effector organs such as glands and muscles. The gap between the axon terminal of one cell and the dendrites of another cell is called the **synapse**. Some neurons are attached to the receptor cells we mentioned earlier, so stimulation of one of our receptors triggers off a train of nervous impulses that passes from neuron to neuron, via the synapses. This transmission is not instantaneous; in humans it is a mere 50 to 100 metres per second, and this partly explains reaction time, the time between our reception of a signal from the outside world and our reaction to it.

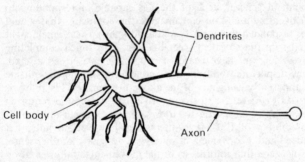

Fig. 3.1. The neuron.

The Brain

The brain itself contains tens of thousands of millions of neurons, each one capable of linking up with about 60 000 others. This is a fantastic communications network.

The brain is split into left and right halves. Fig. 3.2(*a*) shows the **cerebral cortex**, a highly folded plate of neural tissue about 2 mm thick with a total area of about 0.2 square metres. It is the outer crust of the brain, wrapped over and tucked under the cerebral hemispheres. The cortex dominates the human brain and contains about 100 000 neurons in each square millimetre of its surface. Fig. 3.2(*a*) shows the areas responsible for receiving and interpreting signals about smell, sight and hearing.

The sensory and motor regions of the cortex are particularly interesting. An arch that extends roughly from ear to ear across the roof of the brain is the primary **motor cortex** that exercises voluntary control over our muscles. Just behind this and parallel to it is the **sensory cortex**, where signals are received from the skin, the bones, the joints and the muscles. Every part of the motor cortex and of the sensory cortex can be associated with some part of the body. If we were to slice these regions from the top down to the ears we would see a shape like that of Fig. 3.2(*b*). Only one half of each area is shown; the other half would be similar. The right-hand

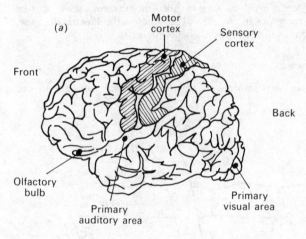

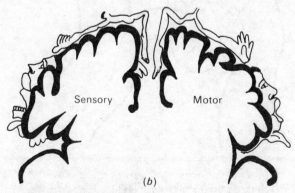

Fig. 3.2. (*a*) The cerebral cortex. (*b*) Sections through the sensory cortex and the motor cortex.

half controls or is stimulated by the left-hand side of the body and vice versa. Systematic surgical exploration of these regions of the brain allows us to identify which parts of the cortex relate to which parts of the body. The distorted human images on Fig. 3.2(b) show this relationship, and you can see, for example, that the brain pays much more attention to sensory signals from the face and hands than from the buttocks and back.

Let us look more closely at the individual senses.

3.3   The Sense of Touch

There are two systems involved. One is called the **cutaneous sense** because stimuli activate receptors in the skin. The other is the **kinesthetic sense**, important to the movement of the body and its parts.

The Cutaneous Sense

Stimulation of the skin informs us of what is directly adjacent to our

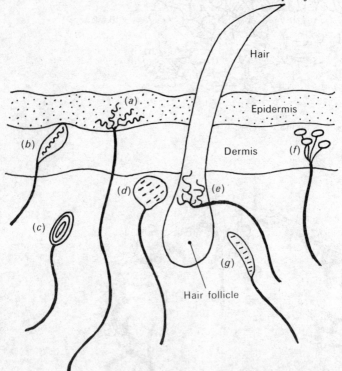

Fig. 3.3. Cutaneous receptors; (*a*) free nerve endings; (*b*) Meissner's corpuscle; (*c*) Pacinian corpuscle; (*d*) Krause's endbulb; (*e*) hair plexus; (*f*) Merkel's discs; (*g*) Ruffini's ending.

bodies. The cutaneous senses make us aware of cold, heat, touch, pressure and pain. It has been proposed that each of these has a different type of receptor.* Heat and cold are detected by **thermoreceptors**, and pressure and pain by **mechanoreceptors**. Fig. 3.3 shows some of the different types. Of the mechanoreceptors, Meissner's corpuscles detect deep pressure; the free nerve endings detect pain. Of the thermoreceptors, Krause's endbulbs detect cold and Ruffini's corpuscles detect warmth.

An important measure of the sensitivity of the skin is the **two-point threshold**, which is the minimum distance by which two pressure-producing stimuli must be separated before we detect two sensations rather than one. The two-point threshold is as low as 3 mm for the index finger, but it is as high as 4 cm for the arms and legs. Try it for yourself! Close your eyes and ask a friend to press two pencil points against your skin. This wide variation in sensitivity of the skin is reflected in the amount of sensory cortex that is devoted to each area of the body (Fig. 3.2(*b*)).

The Kinesthetic Sense

Other types of mechanoreceptors inform us of our movements and of our orientation. In Man each muscle, tendon and joint is equipped with proprioceptors sensitive to muscle tension and stretch. By means of these sensors we are able to perform such tasks as tying a knot with our eyes closed, or of touching our index fingers behind our backs. The muscle spindle is one of the most versatile of our stretch receptors. It detects the stretch of a muscle and Chapter 4 will explain how it helps us to maintain our arm in a fixed position in spite of an increasing load.

A different set of mechanoreceptors signals the body's overall position

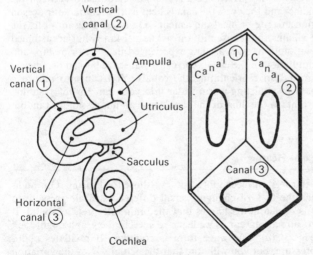

Fig. 3.4. The inner ear signals the body's position in space.

* This simple model is under critical review at present for many instances have demonstrated a lack of one-to-one correspondence between type of receptor and type of stimulus.

in space. They are part of the **inner ear** (Fig. 3.4 and Fig. 3.7). Besides the cochlea (see later), the inner ear consists of two small sacs—the **saccule** and the **utricle**—and three **semicircular canals**. These are organs of balance and posture. Our equilibrium depends upon our sense of vision, stimuli from our proprioceptors, stimuli from pressure sensitive cells on the soles of our feet, as well as upon the inner ear.

The utricle and the saccule signal the position the head is in when at rest. They are small hollow sacs lined with sensitive hair cells and containing small ear stones or otoliths, made of calcium carbonate. As the head tilts, the pull of gravity causes the position of these stones to change, different hair cells are excited and the brain interprets this as a change in the head's position.

The semicircular canals have a different function; they provide a mechanism whereby angular rotations and accelerations of the head can be detected. Each ear has three semicircular canals, each a semicircular tube connected at both ends to the utricle. The canals are so arranged that each one is approximately at right angles to the other two. Each canal contains a viscous fluid that moves with respect to the canal when the head is rotated. This movement of fluid in turn causes a movement of hair cells located in the bulb-like ampulla at the end of each canal. These hair cells are part of the nervous system, so that when they are bent they start a train of impulses on its way to the brain. Since the three canals are located in three different planes, a motion of the head in any direction will excite, at least, the hair cells in one of the canals.

If our utricles and semicircular canals were out of order we could not stand upright with our eyes closed. Without the semicircular canals we could probably stand upright at rest, but as soon as we attempted to move we would be likely to fall over. There is a close relationship between the semicircular canals and the eyes; the canals help us to maintain our eyes in a steady position in spite of head movements. On a less pleasant note, the canals are the villains that cause motion sickness. Man is not accustomed to up and down motion, such as that experienced in a lift or a ship, and the unusual stimulation of the canals can cause nausea. This can be relieved by lying down and so reorientating the canals. You can also induce a feeling of nausea by squirting warm water into your ear, for the resulting convection currents in the fluid of the canals give an impression of random motion.

3.4   The Sense of Hearing

The ear has a third set of mechanoreceptors—those for hearing. Our auditory system has the task of changing small differences in air pressure to varying patterns of neural discharges that the brain interprets as sound. If we plot the pressure at our ear as we listen to speech, the graph might look like that in Fig. 3.5(a). The wave form is complex. It oscillates about atmospheric pressure, but you will note that the actual size of the variation is very small (typically 0.1 N/m^2) in comparison to atmospheric pressure (about $100\,000$ N/m^2). The waveform is complex because it consists of a mixture of different components, at different frequencies and amplitudes.

A typical component is shown in Fig. 3.5(b). Engineers call this a **sine wave** or a sinusoid, and it is the sort of signal that a tuning fork would generate. The two major properties of such a signal are first its **amplitude**, or its maximum deviation from the mean value, and secondly its **frequency**, or the number of times it repeats itself every second. In musical terms, middle C has a frequency of 256 Hz. It has been shown that any complex waveform, such as that of Fig. 3.5(a), can be built up by adding together sinusoidal components with different frequencies and amplitudes. (This is Fourier's Theorem.)

Now this fact considerably simplifies the study of the ear's performance. We can predict its response to very complex wave forms, such as the noise of a jet aircraft or the sound of a cello, by knowing how it responds to the much simpler sine waves (sometimes known as pure tones). So there are two questions we must answer: (1) for sine waves, what range of amplitude

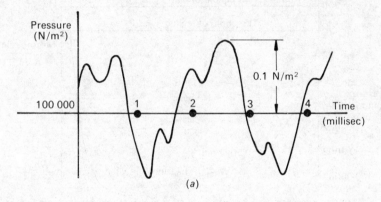

(a)

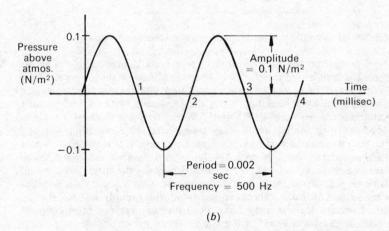

(b)

Fig. 3.5. (a) A typical sound wave. (b) A single sinusoidal component of a wave.

or pressure change can the ear accommodate? (2) for sine waves, what range of frequencies can the ear detect?

Fig. 3.6 supplies the answers to these questions. First frequency: the normal range of human hearing extends from 20 Hz to 20 000 Hz, but usually it is only the very young that can hear at this upper limit. For the majority, sounds above 12 to 15 kHz are inaudible. Fortunately this is more than adequate, for as shown, normal speech covers the frequency range 80 Hz to about 6000 Hz. Those of you with a musical bent will know that doubling the frequency of a note raises it by one octave. Thus one octave above middle C is 512 Hz. So the range of human hearing covers about ten octaves—about ten intervals on the frequency axis of Fig. 3.6. Note that the piano covers just over seven octaves, ranging from 27.5 Hz (A0) to 4186 Hz (C8).

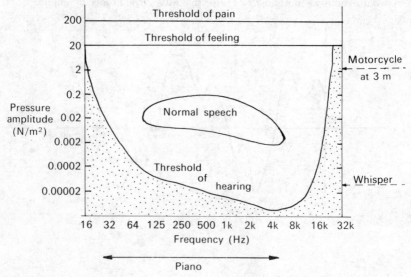

Fig. 3.6. The sensitivity of the ear to pressure amplitude and frequency.

Now turning to the ear's sensitivity to pressure changes, Fig. 3.6 shows three thresholds: the thresholds of hearing, of feeling and of pain. At the higher levels of pressure it has been found that the ear experiences an unpleasant tickling sensation when the amplitude of the sound-pressure variation reaches about 20 N/m². When this is increased to about 200 N/m² (one fivehundredth of an atmosphere), physical pain is felt, and further increase can lead to a permanent deafness. It is interesting to note that, at the threshold of pain, the force on a man's chest and stomach (area 0.5 m²) is about 100 N. At a rocket blast off the spectators are kept far enough away to ensure that this threshold of pain is not exceeded. But even so we can believe the awed spectator who proudly boasted that he could actually feel the rocket taking off. Pressure levels of other common noise sources are included in Fig. 3.6.

Unlike the threshold of feeling and of pain, the **threshold of hearing** varies considerably with frequency. Maximum sensitivity occurs at about 4000 Hz, at which the sharpest ears can detect variations in pressure as low as four millionths of a N/m². It is interesting to note that the frequency of the human alarm call, the scream, lies near this sensitive region at 3000–4000 Hz. Did the ear evolve to match the scream, or did the scream evolve to match the ear?

The ratio of the maximum pressure that we can detect to the minimum is over one millionfold. The ear has evolved to an optimum degree of sensitivity: if it were any higher we would be able to hear the hiss of randomly moving air molecules.

How does the ear operate? Fig. 3.7 shows a sketch of its mechanism. Sound waves are funnelled by the outer ear to the **ear drum** (area 70 mm²), which is caused to vibrate at the frequency of the waves and with an amplitude proportional to the pressure variation of the waves. It has been estimated that this amplitude is only 3×10^{-13}m for a threshold level 3000 Hz tone. This is similar in size to the diameter of a hydrogen molecule.

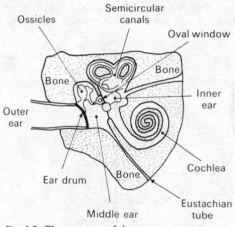

Fig. 3.7. The structure of the ear.

The movements of the ear drum are transmitted by the **ossicles**, a three-bone mechanism in the middle ear, to the **oval window** of the inner ear. You will note that the inner ear is balanced against atmospheric pressure by means of the **eustachian tube** which is connected to the throat. The ear drum is greater in area than the oval window, and this, together with the action of the three-bone lever, produces an eighteenfold amplification of pressure at the inner ear.

The inner ear consists of a fluid-filled bony structure that resembles a snail's shell. It is called the **cochlea**, and its total length is about 40 mm. The motions of the oval window cause waves to move in the fluid of the cochlea, and these are translated into neural charges by some 25 000 nerve endings in the basilar membrane. It seems that the fibres of the basilar membrane are of different lengths along the cochlea, being longer at the tip and shorter at the oval window end. They therefore resemble the strings of a harp or of a piano, the short ones reacting to high frequencies and the long ones to low frequencies.

3.5   The Sense of Vision

The human eye is an extremely sensitive specialised organ for perceiving light. But, just as the ear is restricted to a particular range of intensity and frequency of sound, the eye can only perceive a limited range and intensity of **electromagnetic radiation**.

Electromagnetic waves travel with a velocity of 299 792.458 km/s in a vacuum. Fig. 3.8 shows that the length of such waves, the distance from one peak to the next, varies from the very short 0.001 nm (10^{-12} or one billionth of a metre) to the very long 1000 km (one million metres). The wavelength w, the frequency f and the velocity c of these waves are related by the formula $f = c/w$. So the highest frequency, around 300 million billion Hz, occurs at the short wavelength end of the spectrum, and the lowest frequency, around 300 Hz, occurs at the long wavelength end.

Let us start at the long wavelength end of the spectrum. The longest wavelengths, greater than about 10 cm, are those of the **radio waves**. For

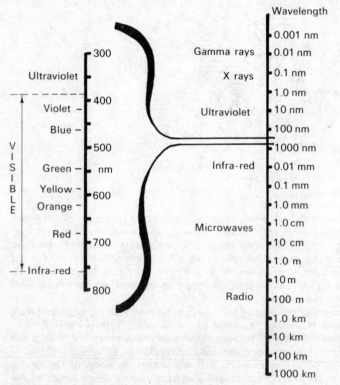

Fig. 3.8. The spectrum of electromagnetic radiation.

example, BBC Radio 4 has a wavelength of 1.5 km and a frequency of 200 kHz. As the wavelength reduces we meet the **microwaves**, used for radar and in microwave ovens. **Infrared** rays, manifested as radiant heat, have even shorter wavelengths. When an object is heated above about 530°C some of the radiant heat has a wavelength of about 750 nm and the human eye detects this as red light: the object is red hot. We have now reached the visible spectrum, a narrow band of wavelengths going from the red end at 750 nm to the violet end at 390 nm. **Ultraviolet** light (1 nm to 390 nm) is outside the visible range, and the temperature must be above 5000°C before any significant amount of this form of radiation can be generated; the sun and the other stars are abundant providers of ultraviolet radiation. Shorter still in wavelength are the **X rays**, generated by firing high-speed electrons at a metal target. And shortest of all are the **gamma rays**, the result of energy changes in the elementary atomic particles that occur in radioactive processes such as nuclear reactors.

Thus, out of this enormous range of electromagnetic radiation, the eye has focused on a tiny portion, that from 390 to 750 nm. In frequency terms the range is from 400 million million Hz (red) to 770 million million Hz (violet).

The eye, like the ear, is extremely sensitive to the intensity of radiation. When it is in the dark for a while it can detect as little as 6 **quanta** of light. This is an astonishingly small amount of light, equivalent to the energy of 6 **photons**, the elementary particles of light. In more familiar terms it is equivalent to the light from a candle 22 km away. The eye can also detect a flash of light lasting only one millionth of a second.

What is the structure of this precision instrument? Fig. 3.9(*a*) below shows a sketch of the eyeball, a finely constructed device that focuses images of objects from the outside world on to the receptor cells located on the back of the eye, the **retina**. Light from a distant object, such as a tree, passes firstly through the transparent **cornea** where it is slightly bent. It then traverses a clear liquid-filled space to the **pupil**, whose size is controlled by the **iris**. The muscles of the iris adjust its size, and hence that of the pupil, depending on the intensity of the light; if it is too bright, for example, the pupil will contract. Having passed through the pupil, the light rays now meet the lens of the eye. The shape of the lens is very important, and the **ciliary muscles** attached to it can adjust this shape according to the distance an object may be from the eye. The lens has to produce a sharply focused image at the retina and you will note, from Fig. 3.9(*b*), that this image is turned upside down and back to front.

The retina is covered by an enormous number of photoreceptors or light-sensitive cells whose function is to detect the light, both in colour and intensity, and to relay this information to the brain. There are two types of cell, **cones** and **rods**, named from their shapes. The rods greatly outnumber the cones, the typical eye having about 125 000 000 rods and 6 500 000 cones. Rods are spread more evenly over the retina. They are more sensitive than cones but they register only black and white. Cones detect colour, but they need plenty of light. They are concentrated in the **fovea**, a small depression in the centre of the retina where images are most likely to fall. Whilst the fovea is the most sensitive area of the retina, the **blind spot**, as

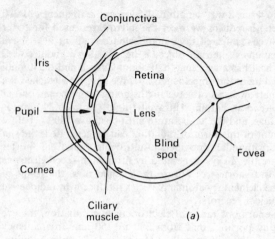

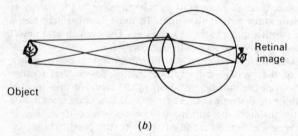

Fig. 3.9. (*a*) The structure of the eye. (*b*) The image on the retina is inverted.

the name implies, is the least. This is the point where the bundle of nerve fibres leaves the eye to form the optic nerve.

One theory of colour vision proposes three different types of cones, sensitive to red, green and blue light respectively, and it is interesting to note that colour television operates on a similar basis. Amongst mammals, only the primates, i.e. the lemurs, apes, monkeys and Man, can see colour. Nevertheless you are advised not to wave any sort of rag at a bull!

Fortunately most of us have two eyes and this gives us stereoscopic vision. Each eye has to turn inwards slightly in order to focus on a nearby object, and we have learned to use this turn as a measure of the distance of the object. It is very difficult to assess distance using only one eye.

Finally a few comments on **acuity**. What is the smallest object that can be seen by the eye? Clearly this will vary from individual to individual. In my own case I am able to see a full stop at 4 m. Now a full stop is about 0.4 mm in diameter, and since the retina is about 15 mm from the lens of the eye, then the size of the image of the full stop on my retina will be about $(0.4 \times 15/4000)$ mm $= (1.50/1000)$ mm $= 1.50\ \mu$m. This is extremely small, and must be approaching the distance between the individual rods and cones. It should be noted that it is not the actual size of the object that determines this limit, rather it is the angle subtended at the eye. For example, knowing that I can see a full stop at 4 m, I could calculate how far away I would have to be from the moon before it just disappeared from my sight. The limiting angle is in my case 0.4/4000 rad. The diameter of the moon is 3476 km, so the distance x can be found from the equation:

$$\frac{0.4}{4000} = \frac{3476}{x}$$

or $\qquad x = (3476 \times 4000/0.4) = 34\ 760\ 000$ km

Compare this with the distance of the moon from the Earth (384 294 km). The astronauts expressed amazement at being able to see the wakes of ships when they were orbiting the Earth. They really shouldn't have been surprised.

3.6 The Senses of Taste and Smell

Chemoreceptors detect odours and tastes. They are sensitive to chemicals that find their way into our nose and mouth.

Our sense of taste is very limited and is probably only intended to distinguish between good and bad food. The average person possesses about 10 000 **taste buds**, located mostly on the tongue but also on the back of the mouth. These taste buds, situated near the tops of the little mounds that you can see on the surface of the tongue, are sensitive to chemicals dissolved in water. (You cannot taste with a dry tongue.)

Four basic tastes have been identified—sour, sweet, salt and bitter—and it has been suggested that a different type of chemoreceptor is associated with each of these. Different areas of the tongue favour different tastes. For example, the tip of the tongue is most sensitive to sweetness, whilst the back is most sensitive to bitterness. The sense of taste is still not properly understood. For example, why should sugar and saccharin taste the same when their chemical structures are so different?

Taste and smell (**olfaction**) are complementary—you know how a stuffy nose can spoil your dinner! Smell, like sight and hearing, allows us to respond to stimulii that are at a distance. It is probably our most primitive sense, but today it is the least important and the least well understood. It is certainly more important for animals who rely on it for detecting food, for recognising each other, for warning against predators and for defence.

Just as light has its three primary colours—red, blue and yellow—from which all other colours can be synthesised, it has been proposed that smell has seven primary odours from which all other smells can be achieved by mixing. These are ethereal, camphoraceous, musky, floral, minty, pungent and putrid. But there is a great deal of debate about this and we are still a long way from a clear understanding of the perception of odours.

The smell sensors are spindle-shaped cells positioned in the mucus film at the top of the nasal cavity. They are stimulated by substances which dissolve in the moist lining of this region. The sense of smell is easily fatigued and a smell experienced for a long period ceases to be noticed. Nevertheless, although not as good as the animals', our sense of smell is still acute, and experiment has shown that we can detect one part of the gas mercaptan in 50 million parts of air.

3.7   Why We need Measuring Instruments

As we have seen, our ability to gather information about the outside world is impressive. Our sensory apparatus has evolved over millions of years to a degree of perfection. But it is not near enough to perfection for the demands of the modern world. Today science and technology, and indeed commerce, require many aspects of the physical world to be determined with high precision. In physics, for example, the masses of the reactive plutonium masses in a nuclear reactor must be known extremely accurately. When manufacturing components on a machine tool, linear dimensions need to be accurate otherwise component pieces will not fit properly. The velocity of an interplanetary probe must be known and controlled accurately. In the chemical industries and the steel-working industries, very high temperatures have to be measured and maintained. In the butcher's shop you will want to be assured that your 1 kg of meat is no less than anyone else's. And so on. Accurate measurement of physical variables is essential today. Without it the scientist could not prove his theories, the technologist could not produce his machines, mechanisms and systems, and business and commerce would be in disarray, because someone would always feel that he had not got a fair exchange for his money.

Why are our senses not up to these tasks? There are several reasons:

(*a*) We have seen that there is a limited range of stimuli to which our sense receptors can respond. For example, we cannot see ultraviolet light; nor can we detect objects less than about one tenth of a millimetre in size. How then can we determine the wavelength of red light which is about 100 times smaller?

(*b*) Many of our senses exhibit a phenomenon called **adaptation**. The neurons signal when a stimulus first appears, but if it remains unvarying the sensing neurons stop signalling. By this means Nature has contrived to make us specially sensitive to change and new events; but this can fool us sometimes. For example, have you ever noticed that the radio, with volume control constant, sounds louder when you get up in the morning than when you went to bed? This is because the ear has adapted itself to extreme quietness during the night. Again, you can demonstrate adaptation to force by standing close to and sideways on to a wall, and pressing your arm against the wall. When you move away after a minute, your arm will feel light and may even rise a few centimetres. A final example involves temperature, and in order to demonstrate it you will need three bowls of water. Fill the left-hand bowl with water at about 35°C, the right-hand bowl with cold water at about 15°C, and the middle bowl with tepid water at about 25°C. Place your left hand in the warm water and your right hand in the cold. When they have both adapted after a minute or two, put both hands in the middle bowl. Your right hand will feel warm and the left hand cool, in spite of the fact they are both in water of the same temperature. So adaptation can lead us to incorrect conclusions about the outside world.

(*c*) The third reason why our senses cannot meet the demands of today is also concerned with our detection of change in the outside world. By how much does a stimulus have to change before we notice the change? Imagine that you are holding a weight of 10 N in your hand. How much additional weight would need to be added before you noticed the difference? This is known as the **'just noticeable difference'** (JND). Experiment has shown that typically the 10 N would have to be increased to 10.2 N before you would spot the difference: the JND is 0.2 N. Now if you were holding 20 N what value would you expect the JND to have? No, it would not remain at 0.2 N. It would increase to 0.4 N. You will notice that 0.2 N is 2 per cent of 10 N and 0.4 N is also 2 per cent of 20 N. The ratio of the JND to the basic stimulus is a constant, a fact discovered by the German physiologist E. H. Weber in 1834. This is known as Weber's Law, and the ratio of the JND to the basic stimulus is known as the Weber fraction. It is listed below for the various senses:

Vision (brightness, white light)	1/60
Kinesthesis (lifted weights)	1/50
Pain (thermal on skin)	1/30
Hearing (middle pitch, moderate loudness)	1/10
Pressure (cutaneous)	1/7
Smell (odour of India rubber)	1/4
Taste (table salt)	1/3

So you can see that we are better at discriminating brightness than weight. Our sense of taste has the least discrimination, requiring a change of 1/3 in the salt concentration before it is noticed. These limitations would preclude the use of human senses from most measurement systems.

(*d*) Finally, we all know that our perception of the signals received by our senses can sometimes cause us to misinterpret the information. This can happen to all our senses, and we touched on this earlier under the heading of adaptation. But the visual illusion is perhaps the one most familiar to all of us. We are easily fooled. In Fig. 3.10(*a*) below, which rectangle is the largest? Check it. Is the cube in Fig. 3.10(*b*) being viewed from above or below? Stare at Fig. 3.10(*c*); you should be able to see things that aren't there! Grey squares should appear at the intersections of white rows and columns. And finally what do you make of the contraption in Fig. 3.10(*d*)?

3.8 The Basics of a Measurement System

We have seen that various properties and shortcomings of our sensory systems make necessary the use of instruments and measurement systems. Lord Kelvin argued the need for measurement when he said: 'When you can measure what you are speaking of and express it in numbers, you know that on which you are discoursing. But if you cannot measure it and express it in numbers, your knowledge is of a very meagre and unsatisfactory kind.'

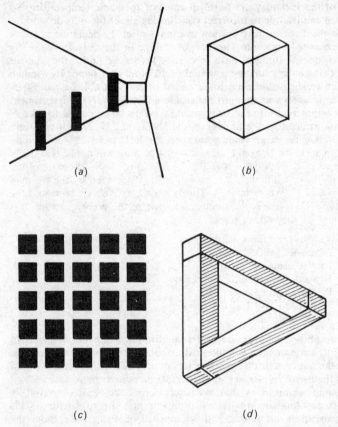

Fig. 3.10. Optical illusions: (*a*) the corridor illusion; (*b*) the Necker cube—an ambiguous figure; (*c*) grey squares that aren't there? (*d*) an impossible object?

What are the basic requirements of a system of measurement? Measurement requires the comparison of an unknown quantity with a similar known quantity. For example, the measurement of length with a ruler is a visual comparison of the unknown length with a scale of known length marked on the ruler. Early units of length were based on the dimensions of the human body. For example, the yard was the distance from the spine to the finger tips when the arms were outstretched. The cubit was the distance from the elbow to the fingertips, varying from 45 to 52 cm. Goliath was 6 cubits and a span in height, between 2.93 and 3.35 m, whilst R. Wadlow, an American who lived in 1918–40, was recorded as 2.72 m.

Now you can see a problem arising with such units. If a small man exchanged 10 yards of his cloth for 10 yards of cloth belonging to a tall man, then the small man would gain from the transaction, for his yards would be smaller than those of the tall man. The way around this is to establish a standard yard, one that everyone would have to refer to, and the first official standard yard was established by Edward I in 1305. It was an iron bar subdivided into 3 feet of 12 inches each.

Basic Units

Nowadays **standards** are extremely precise. During the last few hundred years two measuring systems have been popular: the metric system used by most European countries, and the Imperial system used by Britain and the British Empire. Nowadays these systems are being phased out and most countries are adopting the SI system of units (*Système International d'Unités*). The SI system has six basic units. These are:

the metre	(m)	the unit of length
the kilogram	(kg)	the unit of mass
the second	(s)	the unit of time
the Kelvin	(K)	the unit of temperature
the ampere	(A)	the unit of electrical current
the candela	(cd)	the unit of luminous intensity

There are standards for each of these basic units. Let us look more closely at them.

The metre. The metre was originally defined as one ten millionth of the distance from the equator to the North Pole. A standard metre was made in the form of a platinum iridium bar; it was kept under lock and key in carefully controlled conditions. In 1960 a new, much more precise and unchanging standard of length was introduced. The new standard metre is 1 650 763.73 wavelengths of the orange radiation of krypton-86. (You will recall from Fig. 3.8 that orange light has a wavelength of about 606 nm.)

The kilogram. The primary standard of mass is a platinum iridium cylinder of equal height and diameter. It is kept at the International Bureau of Weights and Measures at Sèvres in France, but duplicates of the standard are held in other countries. The British one is kept at the National Physical Laboratory at Teddington.

The second. The second was originally defined as 1/86 400 of a mean solar day. (86 400 = 24 × 60 × 60). Later in 1955 it was redefined as 1/31 556 925.974 7 of the year 1900. Present demands of extreme accuracy have forced us to use an even more precise definition, based on the atomic clock, and the second is now defined as the duration of 9 192 631 770 periods of oscillation of radiation emitted from a resonating caesium atom.

The Kelvin. The Kelvin was named after the physicist Lord Kelvin, whom we quoted earlier. He predicted that the temperature of −273°C (in practice −273.16°C) was the lowest achievable, since at that temperature there would be no heat energy left in a substance, and all molecular motion would cease. This temperature is used as the zero point for the Kelvin scale. So on this scale water freezes at 273.16 K and boils at sea level at 373.16 K.

The ampere. The primary standard of current was called after the great French scientist Ampère. A magnetic field is generated when an electric current flows in a wire (see Chapter 2), so a magnetic force of attraction or repulsion will exist between two current-carrying wires. The ampere is defined as that steady current which produces a force of 2×10^{-7} Newtons per metre of length between two parallel, infinitely long conductors of negligible cross section, when the conductors are placed 1 m apart in a vacuum.

The candela. The candela replaced the older units such as the international candle. It is defined as the luminous intensity, in the perpendicular direction, of 1/600 000 square metres of a black body which is at the temperature of freezing platinum (2042 K) and under a pressure of 101 325 N/m^2 (an atmosphere).

Derived Units

All other units may be derived from these basic ones, but the unit of force is a particularly interesting derived one, and deserves special attention since many people tend to confuse force, mass and weight. Mass is the amount of matter in a body, and is the same whether the body is on the Earth, on the moon or in outer space. The work of Galileo and Newton led to Newton's famous Second Law which, in simplified form, states that if a force is applied to a body, then its acceleration will be directly proportional to the magnitude of the force and will be in the same direction as the force. In addition, for a given force, if the mass of the body is increased then its acceleration will be reduced by the same factor. These two effects can be combined in the one statement: acceleration a is proportional to the ratio of the applied force F and the mass m of the body, or

$$a = kF/m \qquad (3.1)$$

where k is a constant of proportionality.

This formula makes sense. In essence it says that the harder we push, the greater the effect; and the more massive the object we are pushing the harder we have to push to achieve the same effect. The important point, however, is that it allows us to establish a unit of force. Let us rewrite the formula as

$$F = ma/k \qquad (3.2)$$

There are standard units for both of the variables on the right-hand side of the equation. The kilogram is the standard for mass m. Acceleration a is the rate of change of velocity with time, so its units can be written in terms of standard units as follows

$$a = \frac{\text{velocity}}{\text{time}} = \frac{\text{length/time}}{\text{time}} = \frac{\text{length}}{(\text{time})^2}$$

Now the standard for length is the metre, and the standard for time is the second, therefore the units of acceleration are m/s^2. So we can see from equation (3.2) that the unit for force can be derived from the standard units of mass, length and time. In the SI system the derived unit of force is called the Newton, and is defined as that force which will cause a mass of 1 kg to accelerate at 1 m/s^2. This allows us to put the constant $k = 1$, so that equation (3.2) becomes:

$$F = ma \qquad (3.3)$$

where F is Newton (N), m is kg, and a is m/s².

Equation (3.3) helps us to distinguish mass from weight. We know that the Earth exerts a gravitational force on bodies; and this causes them to fall with an acceleration of about 9.81 m/s². If we use the symbol *g* to denote this acceleration due to gravity, then we can use equation (3.3) to relate weight and mass by the equation:

$$W = mg \qquad (3.4)$$

So our standard kilogram would have a weight of $1 \times 9.81 = 9.81$ N. You will note that, unlike mass, weight can vary with location. The acceleration of gravity on the moon, for example, is about one sixth of that on Earth, so the weight of a body on the moon would be about one sixth of that on the Earth. Even here on Earth, a body's weight will vary with location. Because of variations in *g* it would be more at the poles than at the equator, and it would weigh less at the top of a high mountain than at the bottom of a deep mine.

So you can see how the units for force can be derived from the standard units. See if you can work out how the following units are derived:

Quantity	Unit	Symbol	
Work, energy	joule	J	$=$ Nm
Power	watt	W	$=$ J/S
Electric potential	volt	V	$=$ W/A
Electrical resistance	ohm	Ω	$=$ V/A

3.9 Measurement Systems

In the world of science, technology and commerce there is a multitude of variables that need to be known with some accuracy. We cannot do the subject justice within the confines of this chapter, so it has been necessary to select a few representative areas. These are the measurements of length, velocity, force, time and temperature.

Length and Linear Displacement

We are all familiar with the use of **rules** for measuring length. These can be of wood, plastic or steel. The wooden ones are not so susceptible to heat expansion, but steel ones, on the other hand, are more readily engraved for greater precision and they do not wear so easily. With a good steel rule it is possible to determine a length to within 0.2 mm of its true value.

For greater accuracy it is necessary to use more complicated instruments such as the **micrometer**, which uses the motion of a screw and a nut as its basis of operation (Fig. 3.11(*a*)). An anvil, which presses against the item to be measured, is moved back and forward by means of the rotary motion of a screw thread, and the success of the micrometer depends on the fact that this rotary movement is much greater than the linear movement of the anvil. For example, if the screw thread has a pitch of 0.5 mm then one complete rotation (360°) of the screw would result in a linear movement of only 0.5 mm. If the circumference of the screw head is divided into 50 divisions, then a rotation through one division (7.2°), which is substantial and easily read, corresponds to a linear movement of only 10 μm. If in good condition, a standard micrometer with a total travel of 25 mm should be capable of being read to within 5 μm.

You will note a subtle difference between the steel rule and the micrometer. The steel rule offered a direct comparison of the unknown length with a known one. But the micrometer was less direct in that the linear motion was transformed to a rotary motion. Such transformations are common in measurement, and devices using this principle are known as **transducers**.

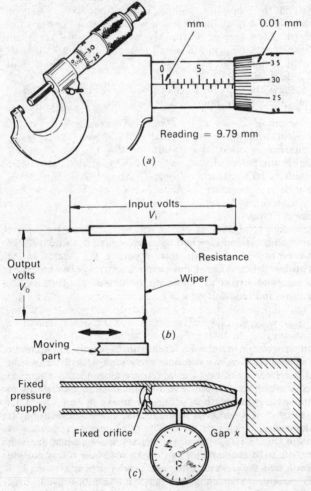

Fig. 3.11. Measurement of length and displacement: (*a*) the micrometer; (*b*) the potentiometer; (*c*) the air gauge.

The **potentiometer**, for example, is a transducer that converts a linear movement into an electrical potential (Fig. 3.11(*b*)). It is a very simple device and finds application in position control systems. It consists basically of a fixed resistor, across which a voltage V_i is applied. A 'wiper' is attached to the moving element and the voltage V_o picked off is proportional to the distance moved.

Another type of displacement transducer transforms linear displacement into pneumatic pressure (Fig. 3.11(*c*)). It is known as the **pneumatic comparator** or air gauge, and its operation depends on the variation of the flow rate of air through an outlet orifice whose effective area varies with the gap *x* between the nozzle and the workpiece. As the gap increases, the outlet orifice area increases, more air can escape and the pressure falls. So a relationship between pressure and gap can be established.

Linear and Angular Velocity

Velocity is the rate of change of displacement, so if we know the distance covered and the time taken, we can find the average velocity by dividing the distance by the time taken. Clearly then the accuracy of the velocity measurement will depend on the accuracy of the measurements of distance and time. For example, the speed of a bullet, or of a racing car, can be found by causing it to pass between a pair of light sources and photoelectric cells a fixed distance, *L*, apart, as shown in Fig. 3.12(*a*). The breaking of the light beams by the bullet can be used to generate voltage pulses which are displayed on one beam of an oscilloscope. If the other beam is connected to an accurate high-frequency oscillator, say at 1 MHz, then the time, *t*, between pulses can be determined to within about $4\mu s$. The average velocity is then L/t.

Aircraft nowadays are nearly as fast as bullets. A bullet from an M16 rifle moves at about 1000 m/s, a speed reached by the Lockheed SR-71A. The Concorde crosses the Atlantic at 600 m/s. But although they may move as quickly as bullets, it clearly would be impracticable to attempt to measure the speeds of aircraft by the method outlined above. The pilot needs a continuous indication of his speed and this is derived from a **pitot static tube** (Fig. 3.12(*b*)). The operation of this device relies on the fact that the pressure of moving air is greater than that of stationary (static) air. The diagram shows that the total pressure of the moving air is detected in tube A, whilst the concentric tube B only detects the static component of the pressure since the holes do not face into the wind. The difference between the total pressure in A and the static pressure in B is the dynamic pressure, which can be shown to be proportional to the square of the velocity. So the velocity can be deduced from the measured pressure difference. The fact that the aircraft, and not the air, is moving, does not affect the argument.

There are many other ways of measuring speed. For example, there are several systems that use the transmission or the reflection of radio waves from the moving object. We shall meet them again in Chapter 8.

The above examples were concerned with linear motion, but many machines involve angular motion. A common means of measuring angular velocity is the **tachogenerator**, which is a small d.c. generator whose arma-

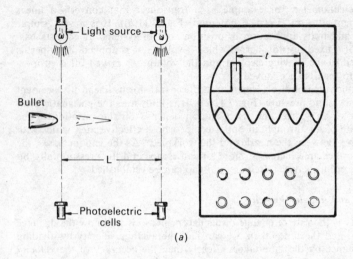

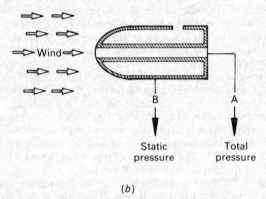

Fig. 3.12. Measurement of velocity: (*a*) by measuring time over a known distance; (*b*) the pitot static tube produces a pressure difference proportional to the square of the velocity.

ture is turned by the shaft whose speed is to be measured. The generated voltage is proportional to the speed of the shaft.

Angular velocity can also be measured by a **strobe flash**, a flashing light source whose frequency can be varied. This technique requires a mark to be placed on the rotating shaft. The flashing frequency is then adjusted until the mark appears to be stationary, and the number of flashes per second will then be equal to the number of revolutions of the shaft per second.

Mass and Force

We emphasised the difference between mass and weight earlier. When you buy sugar, it is the mass, not the weight that you are interested in. A kilogram of sugar is the same on the moon as on the Earth, and you get just as many cups of tea from it; but its weight would be less on the moon. Although there is a clear distinction between mass and weight, it is common practice to determine the mass of an object by comparing its weight with the weight of a standard mass. The **common balance** (Fig. 3.13(a)) below is used for this purpose. The known mass produces a clockwise moment about the pivot of $(Mg)d$, and when the system is in balance the moment produced by the unknown mass, $(mg)d$, is equal to this. Hence in balance $Mgd = mgd$, or $M = m$. The best balances can measure a mass of 1 kg with a resolution of 0.01 mg.

Although more difficult in practice, it is worth noting that mass can also be determined if force and acceleration are known ($m = f/a$). Hence, if a known force were applied to the unknown mass and the resultant acceleration were measured, then Newton's Second Law would allow us to calculate the mass. The known force could be a weight, or in outer space, where weight is negligible, a spring or magnet would have to be used.

Forces, pulls and pushes, including weights, can be measured by using a transducer that converts the force into a linear displacement. The simple coiled spring is the simplest form of such a transducer, and its use in the **spring balance** is illustrated in Fig. 3.13(b). Unfortunately, these direct-reading spring balances are not very accurate since the extensions are relatively small. A different kind of spring, the **proving ring**, is required for high accuracy. The proving ring is basically a high tensile steel circular ring (Fig. 3.13(c)) with loading points situated diametrically opposite, and designed so that the tensile (pull) or compressive (push) loads can be applied. The change in diameter is related to the applied force, and it can be measured by a dial gauge, a micrometer or some other form of displacement transducer. An initial calibration is used to find the relationship between this measurement and known forces.

Strain gauges can be used to determine the deformation of elastic elements such as the proving ring. Such arrangements are known as **load cells**. The strain gauge can be a length of wire or a foil, made up in a grid form and mounted on a nonconducting backing material (Fig. 3.13(d)). This is stuck firmly to the material that is to be tested. When deformation occurs the change in the dimensions of the gauge affects its electrical resistance, and this is used to give a measure of the deformation. Provided that all necessary precautions are taken and the gauge is properly bonded, it is possible to detect a deformation of one millionth of the original length of a test piece. A knowledge of this stretch or shrinkage allows the force to be calculated.

Time

The measurement of time has always intrigued mankind. The earliest methods used (a) the sundial (the time for the sun to traverse the sky), (b) the water clock or clepsydra (the time for water to run out of a vessel),

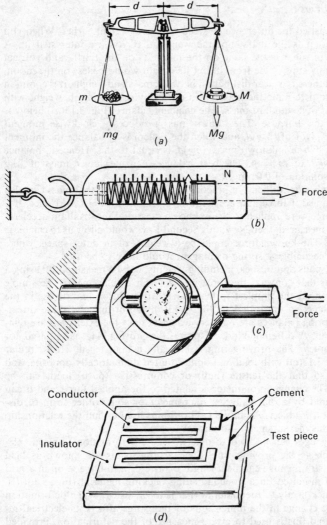

Fig. 3.13. Measurement of mass and force: (*a*) the balance compares masses; (*b*) the
spring balance measures force; (*c*) the proving ring is basically a precision
spring balance; (*d*) a strain gauge.

(*c*) the sand glass (the time for sand to run through a hole) and (*d*) the
graduated candle (the time for a candle to burn away). Clearly these were
not accurate instruments, and the need for greater accuracy led to the
development of the clock.

At the heart of every clock is an **oscillator** that determines its beat and in
turn its accuracy. The **pendulum** was the earliest form of oscillator. In
1583, Galileo, while watching a lamp swinging in an Italian church, timed

its swing and noted that although successive swings became smaller, the time for each swing was the same. In addition, the time for each swing depended on the length of the pendulum. For example, if the distance from the centre of the weight on a simple pendulum to its point of suspension is just less than 1 m, it will take one second for each swing. The Dutch scientist, Huygens, used this principle to make the first pendulum clocks to regulate time accurately.

The pendulum clock, driven by a weight or a spring, counts the number of swings and shows the answer on the clock face. When such a clock is wound up the wheels would race round until the spring ran down, except for the presence of the **escapement**. The escapement wheel releases only one tooth at a time, as shown in Fig. 3.14. You will understand the diagram better if you remember that the escapement wheel has been 'wound up', and the pallets attached to the pendulum allow it to 'escape' once every swing. As shown in (*a*), with the pendulum swinging to the right, the left-hand pallet locks the escapement wheel. When the pendulum starts to the

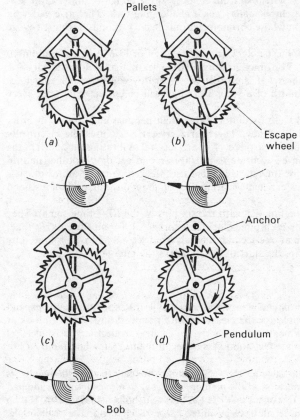

Fig. 3.14. The pendulum clock.

left (*b*), the wheel is freed until position (*c*) when the right-hand pallet locks it again. When it starts to the right again the wheel is again freed and the cycle repeats. So for each swing of the pendulum the escapement wheel makes a move; a move that is used to drive the hands of the clock. And in addition to this the pallets give the pendulum a little push to keep it going. You can see this happening at (*d*) where the point of the tooth nudges the pallet as it slips off it.

The pendulum clock has some shortcomings. Its time of swing is determined by the length of the pendulum, and this is affected by temperature. In addition, the pendulum clock needs a steady base and it is therefore of little use at sea where its regular swing would be upset by the motion of the ship. In 1714 the British Government offered a prize of £20 000 for an accurate means of determining a ship's longitude. This required an accurate clock and since the pendulum clock would not do, attention focused on the use of spiral balance springs. These fine-coiled springs were mounted within wheels which, when set in motion, coiled and uncoiled in a regular fashion creating oscillations akin to those of the pendulum. But springs are also affected by temperature change, so the design of an accurate sea-going clock, the **chronometer**, was a challenging task. The prize was won by John Harrison whose chronometer only lost 54 seconds during a voyage of 156 days.

The typical domestic clock should not gain or lose more than about 10 seconds a day. You may be surprised to learn that a stopped clock is remarkably accurate! It is correct twice per day whilst a clock that loses only one thousandth of a second per day will be correct only once every 118 275 years.

Modern clocks use oscillators other than pendula and springs in order to achieve high accuracy. The **quartz crystal clock** uses the electrically induced vibrations of a piece of quartz crystal. Vibrating at 2.5 MHz the quartz clock can be accurate to a 0.00 1 second per day. Another natural oscillator used for time measurement uses the resonance of atoms of caesium. Such a clock can achieve accuracies of about one millionth of a second per day.

You will all be familiar with the six pips of the BBC time signal. They are controlled by an atomic clock and consist of five pips (of 1 kHz tone) of 0.1 s duration at one-second intervals from second 55 to 59, followed by a longer 0.5 s pip, the start of which marks the minute to $\pm$ 0.05 s.

Temperature

Several different phenomena occur as a result of temperature changes, and they can all be used for the measurement of temperature. We shall look at (*a*) changes in dimensions of a body, (*b*) changes in electrical resistance, (*c*) the thermoelectric effect and (*d*) changes in the radiation emitted by a hot body.

The first group includes the very common **mercury in glass thermometers** in which the expansion of the mercury causes it to move along a graduated scale. Mercury freezes at $-39°C$ (234 K) and boils at 357°C (630 K), so mercury thermometers have a limited range of application. For example, they cannot be used in the coldest regions of the world, such as Vostock in the Antarctica where $-88.3°C$ (184.7 K) has been recorded. In such an environment it is necessary to use an **alcohol thermometer**, for alcohol will not freeze until the temperature drops to $-114°C$ (159 K).

Although mercury boils at $357°C$ (630 K), the range of mercury thermometers can be extended to $510°C$ (783 K) by pressurising the mercury. In such instruments the top end of the thermometer's capillary tube is enlarged into a bulb whose capacity is about 20 times that of the tube. This volume is filled with nitrogen or carbon dioxide at a pressure of 20 atmospheres. You will recall from physics that the boiling point of a liquid increases with pressure, and this phenomenon causes the boiling point of the mercury to rise by about 160 K.

The change in the dimensions of a **bimetallic strip** can also be used for the measurement of temperature. The strip consists of two metals (commonly invar and copper), welded or riveted together so that they must both move in unison. The metals are chosen so that one expands more than the other when subjected to heat. So when heat is applied the difference in expansion causes the strip to bend, and this in turn can be used to move a needle across a scale. We shall see an application for this in Chapter 4.

We now move on to our second category of temperature-measuring devices; those that use changes in electrical resistance. Nickel and copper wire are the two most common materials for **industrial resistance thermometers**, but platinum wire is used for high-precision work. For all three of these metals, an increase in temperature causes an increase in resistance—nickel for example increasing in resistance by a factor of 3 when the temperature increases from $0°C$ (273 K) to $280°C$ (533 K). By using a suitable electrical circuit, such as a bridge, this change in resistance can be indicated on a scale.

These metal resistance elements have the disadvantage that the changes in resistance are relatively small. Greater sensitivity can be obtained by using **thermistors**, made, like transistors, from semiconductors. The sensitivity of the thermistor to temperature change is about 10 times that of the metals. Unlike the metals, its resistance decreases with temperature.

Our third category of instrument uses the thermoelectric effect, discovered by Thomas Seebeck in 1821. He discovered that when a pair of wires of dissimilar materials are joined at both ends to form a circuit, and the two junctions are at different temperatures, then an electromotive force (emf) is generated and a current flows in the circuit. The **thermocouple** uses this Seebeck effect for temperature measurement. Copper and constantan wires are commonly used, giving a voltage output of around 60 mV when the hot junction is at $1000°C$ and the cold at $0°C$.

Our final look at temperature measurement is concerned with the changes in radiation emitted by a hot body. All of the methods we have just discussed involve inserting some kind of sensor into the zone of the measured temperature, and the top limit of temperature is determined by the melting or oxidation of the sensor. Until fairly recently this top temperature has been limited to about $1600°C$, using typical industrial thermocouples. Temperatures above this are measured by the radiation from the hot body. These instruments, called **pyrometers**, can be used in the range $0°$ to $5000°C$. A disappearing filament pyrometer is shown in Fig.

3.15. The tungsten filament is heated electrically until its colour matches that of the measured source, when it becomes invisible to the operator. The current for which this occurs is thus a measure of the temperature and can be calibrated accordingly.

3.10 Errors and Accuracy

Most measurement systems have three basic elements. There is the **detecting element**, or transducer, that converts the quantity being detected into a more convenient form. There is the **signal conditioner** that converts the transducer output into a form that can be displayed. And finally there is the **display unit** that allows the signal to be read and recorded.

Take, for example, the mercury in glass thermometer. In this case the mercury is the transducer, converting the temperature into an increase in volume. The capillary tube is the signal conditioner for it converts the volume change into a distinct linear motion of the measuring column. Finally the display unit consists of the graduated scale.

Errors can creep in at all three of these stages, and the complete elimination of errors for measurement remains impossible except in the most elementary of cases, such as counting the number of fingers a person holds

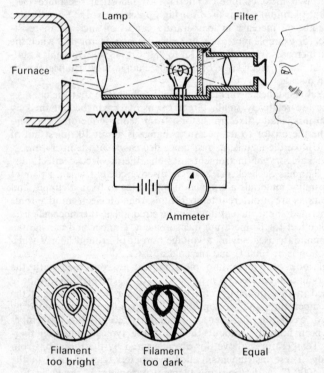

Fig. 3.15. A disappearing filament pyrometer.

up. Having recognised this, it is essential that we have some idea of the magnitude of the errors so that we may know what confidence we can place in our instruments; we must learn to live with error.

Two main categories of error have been identified: systematic errors and random errors—or **accountable** and **unaccountable** errors. I prefer the latter dichotomy. Accountable errors are repeatable and can usually be determined from tests. They can be accounted for! Unaccountable errors are of a random nature, varying from test to test, and we have to make some intelligent attempt at their prediction. Let us look at some examples.

Accountable Errors

There are several sources of accountable errors. Here are a few of them:

(*a*) **Instrument errors.** There may be errors in a scale and in its subdivisions, or the scale may be mounted incorrectly. The needle on a gauge may be bent. The instrument will wear out through use, and friction may increase in its bearings. In addition, the instrument may not be able to respond to rapid changes and this will have to be taken into account.

(*b*) **Interference errors.** When any measurement is made the introduction of the measuring device affects the variable that is to be measured. To take an extreme example, imagine that you tried to measure the temperature of a thimbleful of hot water using a thermometer. A lot of the heat in the water would be lost in heating the thermometer, and the temperature of the water would then be reduced by the presence of the thermometer. The same applies to attempts to measure sociological phenomena. If an interviewer stops you on the street and asks you a question in an attempt to measure public reaction to some event, the very fact that you are being interviewed can affect your views.

(*c*) **Calibration errors.** Calibration is the comparison between one instrument and another more accurate one, under identical conditions, and ultimately this chain of comparison takes us right back to the basic standards we discussed earlier. The importance of knowing the conditions of calibration has to be emphasised. Ideally the conditions in use should be the same as these, but if they are not, the differences should be known so that a correction can be made. For example, a steel rule will only indicate the correct length at a particular temperature.

It is important that the calibration involves reference to a more accurate instrument. I cannot resist a story that illustrates the dangers of ignoring this requirement. Many years ago, when watches were a luxury, a factory inspector asked how the workers knew when to return to work after lunch. He was told that a man on the roof fired a gun at 1.00 p.m. When the inspector asked this man how he knew it was 1.00 p.m. he was told that he checked the time on the clock outside the chemist's shop on the High Street. When the chemist was asked how often he checked his clock, he replied: 'Never, it's always dead right by the one o'clock gun!'

(*d*) **Human errors.** These vary from individual to individual. Some people may read a scale consistently high. It has been shown that many people rarely use the digits 3 or 7 when noting a reading, and this leads to systematic underreading and overreading at these points on the scale.

Unaccountable Errors

Turning now to unaccountable or random errors, we can also list a few examples. Random errors are accidental, small, independent, and arise from many causes:

(*a*) **Mistakes.** Mistakes will always occur. Distractions, tiredness and misunderstandings encourage their occurrence. Reading of scales and the noting down of observations are two areas where mistakes are common. Calculations can also give rise to mistakes. But it is important to note that mistakes are, in general, entirely random and will be just as likely to be high as low.

(*b*) **Changes in environment.** This is another important cause of random errors. Fluctuations in conditions occur randomly. These could include draughts, noises, vibration, magnetic field, frequency of mains supply, temperature, atmospheric pressure. The combination of all these minor influences exhibits itself in a 'scatter' of the measured results. If repeated measurements of a fixed quantity are made, then, even when all of the accountable errors have been accounted for, the measurements will be close to each other, but not the same. These random errors are analysed by statistical methods.

Having now got some idea of the differences between accountable and non-accountable errors, we can go on to distinguish between the **accuracy** and the **precision** of an instrument. Accuracy is the closeness of the reading of an instrument to the true value of the measured quantity. Precision relates to the consistency of the measurement, and a highly precise instrument will have excellent repeatability, with repeated readings of a fixed quantity being in good agreement. It is important to appreciate the difference between these two properties, for it is possible, for example, to have an instrument that is precise but inaccurate. An expensive pressure gauge with a bent needle is such an example. Repeated measurements of a fixed pressure will be in good agreement, but they will all be wrong. In general, we can say that an instrument with small accountable errors will be accurate, whilst an instrument with small unaccountable errors will be precise.

3.11   Communication

Now having spent some time on sensing, it would be appropriate to turn to communication, for the two are related.

Communication is a two-way affair. The outside world communicates information to us, and we in turn, or in response, communicate information to the outside world. We have just discussed the human senses, and it should be clear that they are the means by which the outside world communicates with us. But how do we send information in the other direction? I suppose speech and the written word are the most obvious mechanisms. There are other methods, however, such as gesticulation—have you seen two Italians arguing? And nowadays sociologists tell us that 'body language' is a subtle and unconscious way of informing our fellows of our most intimate thoughts.

Technology has extended Man's ability to communicate to his fellows and machines. For example, it has increased the rate at which words can be communicated between two people. Early Man, by shouting, was able to send about one word per second over a distance of about 2 km. Nowadays, with the help of microwave links, cables and satellites, we can communicate at rates around 10 000 words per second over tens of thousands of kilometres. In addition, technology now allows a single source of information to communicate to a wide audience by radio, television, books, newspapers, video tapes, records, etc.

There were many exciting developments along the way. There was the use of drums, smoke signals and semaphore, but the first big technological advance came with the invention of the telegraph and Morse code (Fig. 3.16) in 1838. **Telegraphy** is a process of signalling by some form of code. The letters and characters of the message are first of all translated at the transmitter into telegraph signals in the appropriate code; these coded signals are then transmitted to the receiver where they are decoded.

A	• —	N	— •	1	• — — — —
B	— • • •	O	— — —	2	• • — — —
C	— • — •	P	• — — •	3	• • • — —
D	— • •	Q	— — • —	4	• • • • —
E	•	R	• — •	5	• • • • •
F	• • — •	S	• • •	6	— • • • •
G	— — •	T	—	7	— — • • •
H	• • • •	U	• • —	8	— — — • •
I	• •	V	• • • —	9	— — — — •
J	• — — —	W	• — —	0	— — — — —
K	— • —	X	— • • —		
L	• — • •	Y	— • — —		
M	— —	Z	— — • •		

Fig. 3.16. Morse code.

Then in 1875 came Alexander Graham Bell's invention, the **telephone**, which permitted speech itself to be transmitted, so avoiding the need to translate the message into some form of code such as puffs of smoke or dots and dashes. The first sentence ever spoken over the telephone was a communication from Bell to his assistant Watson. It was: 'Mr Watson, come here; I want you.' The telephone was a mystery to the man in the street. One attempt to explain it went as follows: 'You know when you stand on a dog's tail it barks. Well if the dog was five miles long it would still bark. That's how the telephone works, only there's no dog.'

The next major advance occurred around 1890 when Marconi and Popov showed that **wireless communication** over long distances was feasible. The Marconi system only transmitted telegraphic messages in the form of pulses of radio waves and it was not until 1906 that a method of transmitting speech was developed. This used the modulation of radio waves, a technique invented by the Canadian physicist Fessender. (We shall discuss this in more detail shortly). Since then there has been an explosive growth in techniques of communication, and today we accept laser beams, fibre optics and satellites as our forerunners accepted drums and smoke signals.

Telecommunications

The name telecommunications is applied to those branches of electrical engineering and physics that are concerned with the communication of information by electricity. Historically, the three main divisions of the subject were telegraphy, telephony and radio transmission. We have already seen that telegraphy handles information in a coded form. Let us briefly examine how telephones and radios use **modulation** to handle information.

We saw earlier in this chapter that human speech has a frequency content varying from about 80 Hz to 4000 Hz. A few sounds, such as the hiss in S, involve frequencies higher than this, perhaps up to 8000 Hz, but if these are ignored the intelligibility of the speech is unimpaired. So when you speak to someone on the telephone or by radio, it is important that the transmission system can handle frequencies up to 4000 Hz. If we wish also to transmit music satisfactorily, it is necessary to cover the full range of audio frequencies (up to 15 000 Hz).

Now there is a problem with transmitting this relatively low frequency range by radio. Although not impossible, it would be quite impractical to have a system that allowed you to talk into a microphone and then transmit the electrical equivalents of the sound waves directly from a radio transmitter. This arises because, roughly speaking, a signal cannot 'get out' of an aerial unless the aerial has a length of at least one quarter of the wavelength of the signal being transmitted. We saw earlier that wavelength w, frequency f and the velocity of light c, are related by the formula $f = c/w$. So the aerial's length has to be at least $0.25w = 0.25c/f$. With $f = 4000$ Hz and $c = 300\,000$ km/s this turns out to be 18.75 km. Clearly it would be inconvenient to have to use such a large aerial.

The way around this is to increase the frequency f of the transmitted signal. This is done by impressing the signal on to a high-frequency '**carrier**' wave, so that the carrier is 'modulated' by the signal (see Fig. 3.17). If, for example, the carrier frequency were 1 MHz, then the required length of aerial would be 75 m, and aerials of this length are relatively easy to construct. At the receiving end the aerial is 'tuned in' to the particular carrier frequency that is being employed, and the received signal is then demodulated to extract the original information.

Fig. 3.18 shows two forms of modulation: (*a*) **amplitude modulation** (a.m.), in which the amplitude of the transmitted signal is modulated by the message, and (*b*) **frequency modulation** (f.m.), in which the frequency of the transmitted signal is the variable. Amplitude modulation is susceptible to **noise** such as that generated by atmospheric static, by electrical machines and even the humble electric shaver. Such noise causes variations in the amplitude of the received signal and the receiver cannot tell if this is part of the original message. Frequency modulation is less sensitive to noise since the receiver looks for variations in frequency and not in amplitude of the signal.

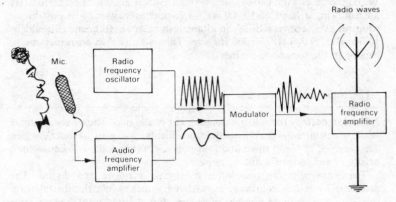

Fig. 3.17. Radio transmission using modulated signals.

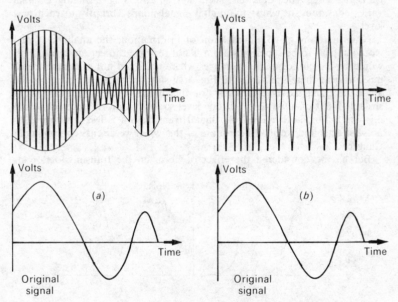

Fig. 3.18. (*a*) Amplitude modulation. (*b*) Frequency modulation.

In the UK, amplitude modulation is used for long; medium- and short-wave broadcasts. VHF (very high frequency) transmissions, with carrier frequencies in excess of 30 MHz, use frequency modulation. The telephone system also takes advantage of amplitude modulation to allow it to cram more and more conversations onto one line. If one pair of wires were needed for every telephone conversation then there would be very long queues, or the country would be a maze of telephone wires. This can be avoided by designating a different carrier frequency to each conversation.

With modern coaxial cables capable of handling frequencies up to 10 MHz and allowing a band of 10 000 Hz for each conversation, it is possible to transmit 1000 conversations simultaneously. Carrier frequencies could be 10 000 Hz, 20 000 Hz, 30 000 Hz, etc. This is known as **frequency multiplexing** in the telephone business.

3.12   Noise and Information

Noise can destroy the information content of a signal. The classic example is the transmitted signal 'Send reinforcements, going to advance', which was received as 'Send three and four pence, going to a dance'. Let us look briefly at methods of reducing noise.

There are basically two kinds of signal: **analogue** and **digital**. The analogue form has continuously variable values, while the digital form can only have a finite number of values (Fig. 3.19). An analogue source can send out a varying signal such as a voice signal, but a digital source transmits a sequence of levels, such as 1, 4, 2, 3, 1. The analogue signal varies continuously whilst the digital signal jumps abruptly to each new level.

Now in terms of communication of information, the analogue signal would appear to be superior since it is not restricted to certain levels. This would be true in ideal circumstances where noise did not affect the transmission. But these are rare, and Fig. 3.19 shows that noise can affect the analogue signal adversely, whilst for the digital signal, it would take a sizeable burst of noise to offset one level so much that the receiver would mistake it for another level. So digital transmission is less susceptible to noise effects, and we shall hear more of this when we discuss computers in Chapter 5.

But having considered the effect of noise on the transmission of in-

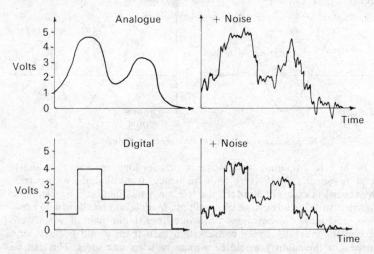

Fig. 3.19. The effect of noise on analogue and digital signals.

formation, it is now appropriate to ask ourselves what we mean by information and how do we measure it? Information is something we gain by reading, listening, or by observing the world. It adds to our stock of knowledge. The technical meaning of information is not radically different from this, but it is more precise.

Shannon published his pioneering work on information theory in 1948. He showed how to measure the amount of information in a message, and how to calculate the ability of a communication to transmit information. As a result we are able to answer such questions as how fast can a computer produce its results? Can a television picture be sent on a telephone line?

How do we measure information? Imagine that I had a box with two compartments, and one of the compartments contained an apple. If you wished to know which compartment held the apple you would need to have information, and you would achieve this by asking questions. How many questions? Only one is necessary. By asking if the apple is in compartment 1, my answer, whether yes or no, would tell you the apple's location. This is the smallest possible amount of information: the choice between two equally likely alternatives such as heads or tails of a coin, or 1 or 0 in a binary number system. It is known as a '**bit**' of information, a word coined from a contraction of 'binary digit'.

Now let us consider a large box with 64 compartments (Fig. 3.20). How many questions need to be asked? In other words, how many bits of information are needed? The answer is 6, and we can demonstrate this using the following series of questions:

1. Is it in the left-hand half of the box?   (No)
2. Is it in the top half of the box?   (No)

This identifies a particular block of 16 compartments.

3. Is it in one of the left-hand 8 of this 16?   (Yes)
4. Is it in one of the top 8 of this 16?   (No)

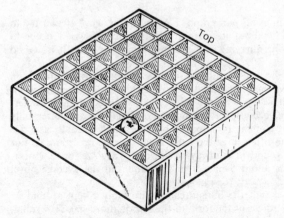

Fig. 3.20. Six yes/no questions must be asked in order to find which compartment is occupied by the apple.

This identifies a particular block of 4 compartments.

     5. Is it in one of the left-hand 2 of this 4?   (No)
     6. Is it in one of the top 2 of this 4?   (Yes)

You will note that there were 64 possible places for the apple to be, and that $64 = 2^6$ (six 2s multiplied together). The power of 2 and the number of questions are the same. This is generally true, so if there are n different possible events, such as an apple in a compartment, then the number of bits B of information required can be found from the formula $n = 2^B$. So if I have a box with 1024 compartments and I tell you where the apple is, I have given you 10 bits of information, since $1024 = 2^{10}$. I have saved you ten questions.

How many bits of information do you receive from a black and white television screen? Each picture can be thought to be made up of $625 \times 800 = 500\,000$ separate points. There are 25 different pictures transmitted each second, so this gives 12.5 million points per second. The brightness of each point can have eight different levels varying from black to white, so since the brightness is a choice from among eight possibilities it represents 3 bits of information ($8 = 2^3$). Hence during each second the 25 pictures include 3×12.5 million $= 37.5$ million bits of information.

One can do similar calculations for other communication media, but let us conclude this chapter by applying information theory to the written word. What is the information content of a word? A letter is selected from an alphabet of 26 symbols plus a space, punctuation marks and so on—let us say a total of 32. Hence a letter has an information content of 5 bits, since $2^5 = 32$. Since the average word has five letters we would conclude that it contains $5 \times 5 = 25$ bits of information. But this is too high, for, unlike the compartments in our box, all letters are not equally likely to occur. The letter e, for example, occurs more often than any other, approximately once in every ten letters, whilst z only occurs once in every thousand.

This is a highly unusual paragraph. Do you know why? If you try to find out what is odd about it too quickly, it probably won't occur to you. Study it without hurrying and you may think of what it is. Good luck.

That paragraph did not include a single e. Most unusual! Based on equal likelihoods we earlier calculated that each letter had an information content of 5 bits – you would have to ask five questions to determine it. In practice, because of the nature of English, each letter only has an information content of about 1 bit. We don't need to ask five questions for we can often infer the letter from the one preceding it. For example u follows q, and th is always followed by a vowel or r, w or y. Thus a word carries about 5 bits of information and the average novel about 250 000 bits.

It has been shown that the human can comfortably accept around 25 bits per second, regardless of the form of the information, speech, writing, etc. Computers are much better in this respect, as we shall see in Chapter 5.

3.13   Exercises

1. (*a*) Describe the four major types of receptors to be found in the human body, giving examples of each. (*b*) Explain the process of transmission of information from these receptors to the central nervous system.

2. (*a*) Name the six senses. (*b*) Write 200 words about the mechanisms of the ear. (*c*) What is the threshold of hearing? (*d*) At what frequency is the ear most sensitive?

3. (*a*) Sketch the electromagnetic spectrum, indicating the frequencies of X rays, radio and visible light. (*b*) Make a sketch of the human eyeball showing the major elements. (*c*) How is colour perceived?

4. Describe four shortcomings of our sensory systems that required Man to resort to measurement systems.

5. (*a*) Why are standards necessary? (*b*) What are the standards of mass, length and time? (*c*) Show how the unit of force can be derived from these standard units. (*d*) Explain why the volt is a derived unit.

6. The year is 1984. Orwell's 'Big Brother' wishes to control the love, loyalty, hate, diligence and motivation of the people. Of course, in order to do this he needs some measure of these variables. Suggest possible standard units.

7. Define the following terms: transducer, signal conditioner, display element. Identify these elements in a spring balance and in a pendulum clock.

8. Describe the operation of transducers for relating (*a*) displacement to voltage, (*b*) strain to resistance, (*c*) temperature to resistance, (*d*) speed to pressure, (*e*) displacement to pressure.

9. (*a*) What are the relative advantages and disadvantages of analogue and digital systems? (*b*) Describe an analogue method and a digital method for measuring the speed of rotation of a shaft.

10. (*a*) Describe three sources of accountable errors and three sources of unaccountable errors in measurement systems. (*b*) Distinguish between accuracy and precision.

11. (*a*) Why is modulation used in radio broadcasting? (*b*) Distinguish between amplitude modulation and frequency modulation. (*c*) What is frequency multiplexing?

12. (*a*) What is a bit of information? (*b*) How much information do you receive (i) if you are told the location of a diamond in a safe with 128 locked compartments? (ii) if you are further told that the code required to open the compartment is cxgtm? (iii) if instead, you are told that the code is 53176? (*c*) How much information is contained in question 4 of these exercises?

3.14   Further Reading

Those marked with an asterisk are particularly recommended.

Attneave, F., *Applications of Information Theory to Psychology*, Holt, Rinehart & Winston, 1959.
Barrass, R., *Human Biology Made Simple*, Heinemann, 1981.
Bolton, W., *Engineering Instrumentation and Control*, Butterworths, 1980.
*Brinkworth, B. J., *An Introduction to Experimentation*, English University Press, 1973.
Fisher, R. B., *Brain Games*, Fontana, 1981.
*Gleitman, H., *Psychology*, W. W. Norton & Co., 1981.
*Glorioso, R. M., and Francis, S. H., *Introduction to Engineering*, Prentice Hall, 1975.
Mackean, D. G., *Introduction to Biology*, John Murray, 1973.

Mind and Behaviour, W. H. Freeman (reprints from *Scientific American*), 1980.

Penny, R. K., *The Experimental Method*, Longman, 1974.

Smith, B. J., Peters, R. J., and Owen, S., *Acoustics and Noise Control*, Longman, 1982.

Sperling, A., and Martin, K., *Psychology Made Simple*, Heinemann, 1982.

The Human Component, Speech, Communication and Coding, The Open University, T100, 2 and 3, 1971.

The Man-made World, Polytechnic Institute of Brooklyn, 1968.

Villee, C. A., and Dethier, V. G., *Biological Principles and Processes*, W. B. Saunders Co., 1976.

Young, J. F., *Information Theory*, Butterworth, 1971.

4
Technology Extends Man's Control

4.1 Introduction

Human beings are always exercising control, whether consciously or subconsciously. At the subconscious level, control systems ensure that the body temperature is at the right level, that the heart is beating at the correct rate, that blood pressure is satisfactory, and so on. At the conscious level every task we perform requires us to exercise control. As I write this I am controlling the position of my pen to form these words. When I speak I have to control the position of my lips, my vocal chords, my tongue, and the pressure of the air in my lungs in order to make an intelligible sound. When I lift an egg I have to control the position of my hand and the force exerted by my fingers.

With the advent of engines and machines, Man soon discovered that his control abilities were being exercised to the limit. Driving a car, for example, is a complicated task, but one that is within our capabilities. It requires the control of position and velocity. On the other hand, flying a modern airliner, or a vertical take-off aircraft, or indeed a lunar module, is a much more difficult task—one that exceeds the control capabilities of most people. In such cases technology has provided means of extending our capabilities, for although the human system has evolved to a high degree of perfection, Nature never envisaged that it would have to control the sort of systems we have just mentioned. Some of our shortcomings are (*a*) a delay of about 0.25 seconds between deciding and acting, (*b*) a reduction of performance with fatigue and (*c*) an inability to work in extreme environments.

These human limitations have become particularly significant over the past 40 years, mainly because of the ever increasing demands for accuracy of control. Nevertheless, it was the ancient Greeks who first appreciated the significance of control in human affairs, and their word for a ship's helmsman is the root for the modern word **'cybernetics'**, defined by Norbert Weiner as 'control and communication in the animal and the machine'. The helmsman was the archetypal cybernetician, for his task of steering the ship required the communication of information—he had to be informed of where he was and where he was supposed to be going. We shall hear more of this later in the chapter.

4.2   Amplifiers

Amplifiers play a major role in control systems. They provide a means of controlling the amount of energy available, and they are called amplifiers because they use small amounts of control energy to produce large energy flows. Fig. 4.1 illustrates this concept. Take a domestic electric switch as an example of an amplifier. The control input in this case is the energy that is needed to flip the switch; the external energy source is the electrical power available in the mains; the load on the system could be a washing machine, an electric fire or a food mixer. The **switch** is an amplifier because it uses a small amount of mechanical energy to control a large amount of electrical energy.

4.3   On-off Amplifiers

Besides being an amplifier, the switch has another peculiar characteristic. It has only two states, i.e. it can only turn the power on or off. For this reason it is referred to as an on-off amplifier. A **clutch** is an example of a mechanical on-off amplifier (Fig. 4.2). It either does or does not transmit the full power from the engine shaft to the gearbox (neglecting slip)—so it is an on-off device. And it is also an amplifier, for the input energy from the driver's foot is negligible compared to the power generated by the engine. In the commonest form of clutch, the friction between two or more

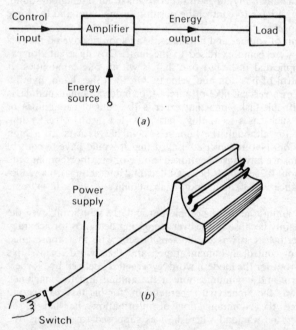

Fig. 4.1. (*a*) Block diagram of an amplifier. (*b*) The switch as an on-off amplifier.

discs is used to transmit power. When the clutch is engaged strong springs hold the discs together, and for disengagement a pedal is used to force the discs apart. It is interesting to note that some early cars did not have a clutch, and the aptly named 'crash' gearbox was used to connect the engine shaft to the drive shaft, and for changing gear. In those cases, when going from neutral to drive, the gearbox itself acted as an on-off amplifier.

There are many examples of on-off amplifiers in the field of fluid power transmission. A very elementary one was demonstrated by the early mill-

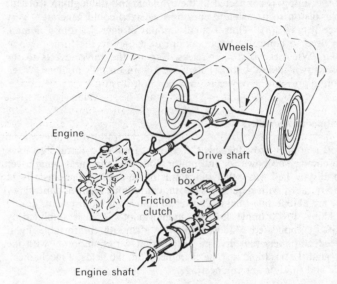

Fig. 4.2. The clutch is an on-off amplifier assuming it is not allowed to slip.

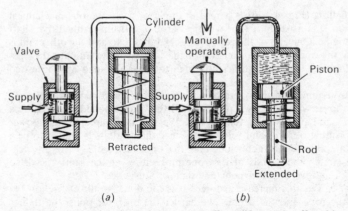

Fig. 4.3. A pneumatic valve acting as an on-off amplifier: (*a*) in off position; (*b*) in on position.

wright when he opened the sluice gate to allow the water to rush to his waterwheel. To turn off the power at his wheel all he had to do was to close the gate again.

A modern equivalent of turning on and off a stream of fluid is given in Fig. 4.3, which shows how a **pneumatic valve** controls the air flow to a pneumatic cylinder. In Fig. 4.3(*a*) the valve is shown in its closed position, in which the supply from the air compressor is isolated from the line connecting the valve to the cylinder. In this condition a spring holds the piston and rod in the retracted position. When the valve button is pushed down by hand the supply is connected to the cylinder, and the high-pressure air pushes the piston to its extended position where it could generate a very large force (Fig. 4.3(*b*)). Thus a small amount of energy, from a human, has been used to control a large amount of energy in the form of compressed air. When the button is released again, the valve reverts to the condition shown in Fig 4.3(*a*). This is clearly an on-off amplifier. When the button is pressed the power is on; when not, it is off.

4.4 Continuous Amplifiers

The on-off amplifier provides only the coarsest form of control. You have only two choices: no power or full power. Can you imagine driving a car in which all you had was either full power or no power? The majority of tasks we are faced with require a smooth, continuous form of control, by which we are able to fine-tune the power to match our requirements at any instant. Think, for example, how the human pilot's input is amplified by an aircraft's complicated systems to provide a smooth comfortable flight. So we need amplifiers that give an output that varies smoothly with the control signal. Let us look at some examples from the fields of mechanical, electronic and fluid power engineering.

A Mechanical Amplifier

The **windlass** (Fig. 4.4(*a*)) is a nice example of a continuous mechanical amplifier. It consists of a rope wrapped around a drum on the drive shaft of an engine. One end of the rope is attached to a load and the other end is held by the operator. The function of the windlass is to amplify the operator's energy to a level sufficiently high to move the load at the desired speed.

Let us assume that the operator pulls his end of the rope with a force T_1 and it moves with velocity V_1; as a result the load is pulled upwards by a force T_2 and moves with velocity V_2. We saw in Chapter 2 that power is the product of force and velocity, so since the windlass is known to be a power amplifier, we can conclude that $T_2 V_2$ must exceed $T_1 V_1$. Now, neglecting stretch, both ends of the rope must move at the same speed; so $V_1 = V_2$ and we therefore conclude that T_2 must exceed T_1.

What causes this increase in force? It is the direct result of friction between the rope and the drum. The harder the operator pulls, the tighter the rope grips the drum and the more power it can take from the engine. So we expect T_2 to increase with T_1. Look at the extreme cases. When

$T_1 = 0$, the rope is slack and the drum turns easily inside the coils; no force T_2 is developed and no power is taken from the engine. The other extreme is the case where the force T_1 is very high and the resultant friction force does not allow any slip between the rope and the drum. Here we may as well assume that the rope is attached to the drum, in which case it is clear that all of the engine power will be delivered to the load. So when T_1 is zero, no power is transferred to the load, and when T_1 is very large all the engine power is transferred. For values of T_1 between these two extremes, it can be shown that the output force T_2 is proportional to the input force T_1. In a practical windlass, with about one and a half turns of rope, T_2 will be about 100 times T_1.

The characteristics of the windlass amplifier are sketched in Fig. 4.4(*b*), which is typical of many types of amplifier (except that in this case negative pulls (pushes) cannot be amplified). The figure shows a region of continuous control. It also illustrates the phenomenon of **saturation**, which is an

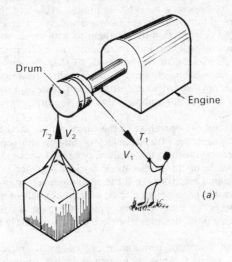

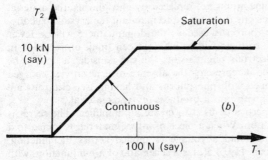

Fig. 4.4. (*a*) The windlass can act as a continuous amplifier. (*b*) Relationship between output pull and input pull.

unavoidable feature of all amplifiers. You cannot go on amplifying indefinitely, for at some stage you will find that you are asking for more energy than the power source can supply. When an on-off amplifier is used, such as a light switch, there is no region of continuous control, and the output can only adopt the two states, zero or saturation.

Electronic Amplifiers

The electronic amplifier is a very popular device. It is found in most homes in record players, televisions and radios, and it is predominant in industry and commerce. The earliest electronic amplifiers were **thermionic** devices, relying on the fact that certain metals emitted electrons when heated, but these were on the whole fairly large and delicate components, a typical triode being an evacuated glass cylinder about 5 cm in length and 2 cm in diameter. Nowadays, solid state microelectronics are taking over and the **transistor**, invented by Shockley in 1960, is rapidly replacing the older thermionic amplifiers. Transistors have many advantages. For example, their small and rugged construction allows them to withstand mechanical shocks and vibration that would ruin a thermionic valve. Their smallness also allows density of packing to be increased, so that a complete electronic product can now be much smaller than earlier versions. Again heating filaments are not needed in transistors and there is therefore a saving in power, for a small thermionic valve can require as much as 2 watts for this purpose.

In order to understand the operation of the transistor we need to know something about the properties of **semiconductors**. Silicon and germanium are the two main semiconducting materials. Crystals of these materials, and of metals, are made up of atomic nuclei locked together in a fixed lattice by their associated electrons. In a metal like copper, many of the electrons can move easily, so, since an electric current consists of a flow of electrons, it is easy to set up a current in copper—it is a good conductor. In pure silicon or germanium, however, the electrons are relatively fixed, so they are poor conductors.

The conductivity of silicon or germanium can be greatly increased by the introduction of impurities, a process known as **doping**. Impurity atoms do not fit neatly into the regular atomic structure of pure silicon or germanium crystals. If some atoms are replaced by phosphorus, the material ends up with more electrons than needed to maintain the crystal structure. If, on the other hand, boron is used as the dopant, there will be fewer electrons than needed, so there will be what we can think of as 'holes' in the structure where electrons ought to be. We can consider a hole to be positively charged since it represents the absence of a negatively charged electron. Materials doped with phosphorus and having free electrons are called 'n' type. Doping with boron produces 'p' type material.

How can these properties be used to create an amplifier? The secret is the use of the *p-n* junction. When a piece of *n*-type material is joined to a piece of *p*-type material, and a voltage is applied across the junction, interesting things happen. Fig. 4.5(*a*) is a schematic of a *p-n* junction, with free electrons shown as dots, and holes as circles. If a battery is applied across the junction, with negative terminal connected to the *p*-type material

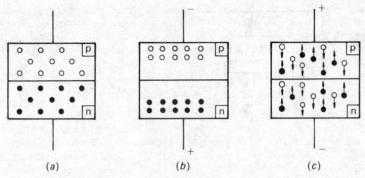

Fig. 4.5. A *p-n* junction as a diode: (*a*) free state; (*b*) positive voltage applied to the *n*-type material, no current; (*c*) positive voltage applied to the *p*-type material, current flows.

and the positive terminal to the *n*-type material, then the electrons will be attracted to the positive terminal and they will move away from the junction (Fig. 4.5(*b*)). Similarly, the positively charged holes will be attracted to the negative terminal. No charge crosses the junction and so no current flows. Now, if we reverse the battery connections, a completely different situation results (Fig. 4.5(*c*)). This time the electrons will be attracted across the junction to the positive terminal and the holes will flow the other way towards the negative terminal. A current has been established, but since the *p-n* junction only conducts in one direction it is known as a **diode**.

A transistor consists of two of these diodes, back to back. Fig. 4.6(*a*) will help us to understand the operation of the *n-p-n* transistor. Note that the bottom piece of *n*-type material is called the **emitter**, and the top portion the **collector**. The **base** layer is *p*-type and is usually extremely thin. With the switch in the open position no current can flow through the transistor since there are two diodes back to back, opposing each other. When the switch is closed, as shown, the bottom *n-p* junction is in a similar state to that of Fig. 4.5(*c*). Electrons can now flow from the emitter to the base, and having reached the base they can either flow towards the positive terminal by way of the resistor in the base circuit, or through the collector. The resistor in the base circuit inhibits the flow of electrons along that route, so the bulk of the electrons passes through the *n*-type collector which is rich in electrons, all being attracted to the positive terminal, and all the while being replaced by the flow of electrons from the emitter.

Now we begin to see the basis of the amplification process. A small flow of electrons through the base circuit is associated with a large flow through the collector. In other words, a small base current can give rise to a large collector current. With proper adjustment of the resistor in Fig. 4.6(*a*) it is possible to achieve current amplification of as high as 150. For example, typically, a change of base current of 10 μA could cause a 1.5 mA change in the collector current. Fig. 4.6(*b*) shows one arrangement of a transistor amplifier used for stepping up an input voltage. Unlike the transistor of Fig. 4.6(*a*), which operated in an on-off mode (switch open or closed), this circuit is designed to amplify a continuously varying input and to produce

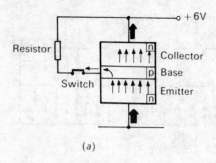

(a)

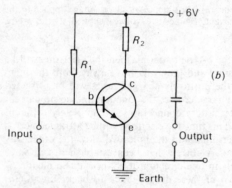

(b)

Fig. 4.6. (a) The *n-p-n* transistor. (b) An *n-p-n* transistor used for voltage amplification.

an output with similar variations, but larger. There is not enough space to allow a detailed discussion of the amplifier here, but it is hoped that this brief introduction to the mysteries of the semiconductor will help to dispel fears of the unknown.

Fluid Amplifiers

The role of **fluidics** in fluid power is similar to that of electronics in the field of electrical power. The success of electronics encouraged engineers to look at the possibility of using fluid streams instead of electron streams for amplification purposes. Such fluid devices can have advantages over their electronic counterparts, such as immunity to radiation and extremes of temperature. However, although being small, they cannot match the minuteness of microelectronic components. And since electrons move at the speed of light, fluid elements cannot match the switching speeds of electronic devices. Nevertheless, the special properties of fluidic elements have resulted in successful applications in the aerospace field, in nuclear reactors and in the control of pumps for handling the extremely cold liquid gases. The earliest work was carried out in 1959 at the Diamond Ordnance Fuse Laboratories and at the Massachusetts Institute of Technology in the USA.

A **continuous fluidic amplifier**, which behaves in an analogous fashion to the transistor, is sketched in Fig. 4.7(*a*). It consists of a supply tube, a receiver tube and an input or control tube. Variations in input signal deflect the main supply jet, hence changing the output signal (pressure, flow or power) in the receiver. The device acts as an amplifier since changes in the input can cause larger changes in the output. For example, good design can result in a change in output pressure up to 20 times the change in input pressure. These characteristics are shown in Fig. 4.7(*b*), and you should note a similarity to the characteristics of the capstan amplifier shown in Fig. 4.4(*b*). This time, however, the graph is the other way around for an increase in the fluidic amplifier input causes a reduction in output. Because of this, the fluidic amplifier is said to have a negative gain. A typical fluidic amplifier would be about 2 cm by 3 cm in size, and would use air at very low pressure, around one tenth of atmospheric pressure.

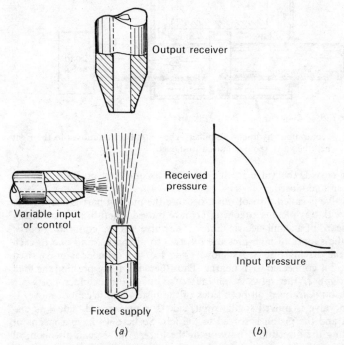

Fig. 4.7. (*a*) A continuous fluidic amplifier. (*b*) Output pressure decreases with input pressure.

Much higher pressures and power can be handled by the **hydraulic amplifier** shown in Fig. 4.8. It uses a valve to supply pressurised oil to a piston and cylinder which in turn drives a load. A comparison of this continuous amplifier with the on-off pneumatic amplifier of Fig. 4.3 should reveal some similarities, for it too uses a valve and cylinder. However, whereas the pneumatic valve was designed to have only two operating conditions,

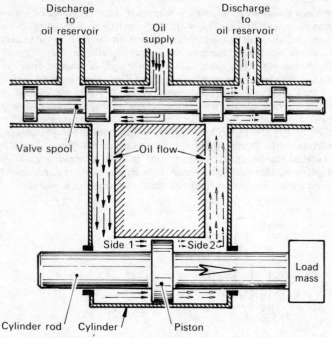

Fig. 4.8. A continuous hydraulic amplifier. The output piston moves to the right
   when the valve spool is moved to the left.

open or closed, the valve in this case can assume any position between
fully open and closed.

The valve is called a **spool valve**, because the moving part is shaped like
a spool with tiny pistons attached. It can be moved to left or right manually
or by means of a small electric motor. The valve has five connecting ports:
one to the supply of high-pressure oil, two to discharge, and one to each
side of the driving piston. As shown, side 1 of the cylinder is pressurised
and side 2 is connected to discharge. Thus there is a force pushing the load
to the right. If the valve is centralised so that both cylinder ports are
blocked, oil is trapped on both sides of the piston and it cannot move. If
now the valve is moved to the right, side 2 is pressurised, side 1 is dis-
charged and the piston is forced to the left. So here we have a means of
controlling the direction of motion of the load. And we can also control
the velocity of the load for this depends on how far the valve has been
moved. The greater the opening of the cylinder ports, the more oil can
flow through and the faster the piston will move. Indeed, this is what
makes it a continuous amplifier rather than an on-off device.

In the hydraulic amplifier, the input power is that required to overcome
the friction and fluid forces acting on the valve spool. These are usually
small, less than 20 N. The output power can be considerable, typically
50 kW, and output forces of 10 kN are quite normal. So the hydraulic
amplifier is popular in applications that require large forces, such as power

steering systems for automobiles, operating ship steering gear, driving the control surfaces on aeroplanes, positioning the cutters on machine tools, and so on.

4.5   Open Loop Control

We now have some idea of how the basic energy amplifiers work. Let us take a closer look at one of them—the switch and electric fire combination. The function of an electric fire is to provide a known amount of heat, and when we buy a 1 kilowatt fire we assume that it will convert that amount of electrical power into heat. So the switch and fire combination is a form of control system. Flipping the switch produces a controlled amount of heat. But where does the control come from? How does the fire know that it is expected to convert 1 kilowatt of electrical power to heat? Engineers have selected a particular type of resistive element for the fire, so that when it is connected to a 240 volt, 50 Hz supply, it will dissipate 1 kW of electrical energy. The fire can do that job and no other. Its one aim in life is to dissipate 1 kilowatt of electrical power, and the engineers have helped it to meet that aim by designing it properly. So, going back to our earlier question—where does the control come from?—we should now be able to see that, for the electric fire, its design is its control.

Now imagine that you have such a fire in your greenhouse, operated by a switch in your living room, and you have learned from experience that the plants will flourish if you switch it on for an hour on winter evenings. When you flip the switch in the living room you are confident that 1 kilowatt of energy is at work in the greenhouse. But can you be sure? What would happen if the supply voltage fell due to overload? Clearly you would get less heat than you expected, and your plants would suffer. The control system would have failed, and you would be sitting in the living room happily unaware of the disaster in the greenhouse. Why did it fail? It failed because the engineers designed the fire under certain **calibration conditions**, and these conditions had not been met in the greenhouse. The calibration conditions were voltage 240 V, frequency 50 Hz, the greenhouse conditions were voltage 100 V, frequency 50 Hz.

A similar catastrophe could confront you if the washing machine voltage was less than the designer's calibration value. The motor would not run as quickly as required and at the end of the set time your greasy overalls would still be a long way from the pristine state.

These were examples of on-off systems not doing their jobs properly. Let us now look at a continuous amplifier—one for controlling the speed of a steam turbine. The system is shown in Fig. 4.9. The amount of steam flowing to the turbine is controlled by the needle valve. When the valve is wide open, the turbine runs at maximum speed and the kinetic energy of the load is at its greatest. As the valve is progressively closed, the speed will reduce and the turbine will ultimately stop when the valve is fully closed. Now, in designing such a system the designer would need to know what load had to be driven and what steam pressure was available. Having used his knowledge of engineering science to design the system, he would then test the finished product in the laboratory before despatching it to the customer. During these tests he would maintain certain calibration conditions; the steam pressure would be held at a constant value, and the load would be of a known fixed amount. He would then measure the turbine speed for various valve settings and draw a **calibration curve** like the one shown in Fig. 4.9(*b*). The designer has now produced a system that is controlled by its design to produce a certain speed if certain conditions are met.

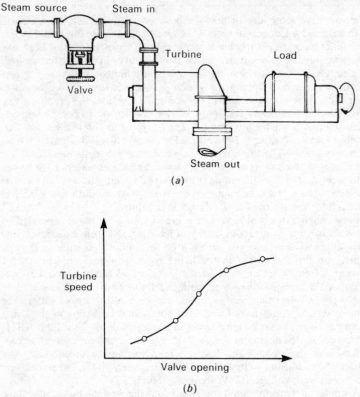

(*a*)

(*b*)

Fig. 4.9. (*a*) An open loop speed control system. (*b*) Measured characteristics under calibration conditions.

In due course the machine is delivered to the customer who installs it, ensuring that the steam pressure and the load are those that the manufacturer has assumed. He now wants it to run at a certain speed, so running his finger along the calibration curve that the manufacturer has supplied, he notes the valve opening that is required for that particular speed. He adjusts the valve to that setting and then goes off to do some other job, happily assuming that all the machines drive by his turbine are now running at the right speed. But can he be sure? What happens if the steam pressure falls, or if friction increases the load? Once again, like the on-off electric fire system, this continuous system fails to produce satisfactory control of speed because the calibration conditions are not being met.

What can we do about it? What is the matter? Basically there is a short-age of information. Too many assumptions are being made: the operating conditions are assumed to be the same as the calibration conditions; the output is assumed to be at the desired level. One possible way around this is to provide more information by continually measuring the operating conditions and adjusting them to equal the calibration conditions. But this in itself would require further control systems—a system to ensure, for example, that the steam pressure always stayed at the design value. Surely the most direct and most obvious action is to gather information about the thing that we are actually attempting to control. If we measure the output speed of the turbine then we will know precisely how our control system is performing. We will not have to make any assumptions!

4.6 Closed Loop Control

This is where **feedback** comes to the designer's aid. Information concerning the output of a system is fed back and appropriate corrective action is taken. This is illustrated in Fig. 4.10, which shows how a human operator

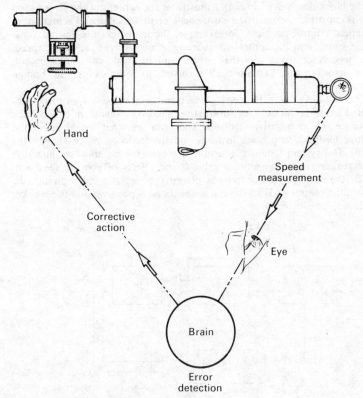

Fig. 4.10. Closed loop speed control system.

can provide the feedback in our speed control example. First an instrument is required to measure the actual speed of the turbine. The operator notes the output speed indicated by this instrument, determines the error between it and the desired output speed, and takes corrective action. For example, if the output speed is 900 rev/min, and the operator wants 1000 rev/min, he subtracts the actual output from the desired value and calculates an error of 100 rev/min. This indicates that the turbine speed is too low, so the appropriate corrective action is to open the steam valve wider. This will cause the turbine to speed up and the operator will repeat these procedures again—noting the speed, determining the error, adjusting the steam valve. In due course, the skilled operator will be able to set the output speed at the desired value. Of course, in reality the whole process is continuous and the operator does not consciously calculate the precise error. His task is to make the pointer on the speed measuring instrument line up with the desired value.

Because of the flow of information from output to input, feedback systems are referred to as **'closed loop systems'**. Systems without feedback, such as the greenhouse heating system, the washing machine and our original speed control system, are called **'open loop systems'**.

The block diagram of Fig. 4.11 illustrates the concept of the closed loop and its constituent elements, **measurement, error detection** and **actuation.** In the speed control problem, for example, the input was the desired value of the speed, and the controlled quantity or output was the actual speed. The process or plant was the turbine and its load, and measurement, error detection and actuation all involved, to some extent, the human operator.

You will have noted that error detection requires the output to be subtracted from the input. This subtraction has led to this form of feedback being known as **negative feedback.** Can you see what would happen if **positive feedback** were used instead, and the feedback was added to the input? The system would run away. It would be unstable (see later). A loudspeaker system sometimes exhibits the effects of positive feedback. When the speaker (the output) is placed too near the microphone (the input), the output adds to the input, feedback is positive, and the speaker emits a howl of protest.

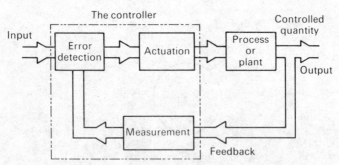

Fig. 4.11. Block diagram of a general closed loop system.

4.7 Control and Human Operators

I am sure you can think of many other control systems that involve human operators. Driving a car is one example. There are two controlled quantities here: the speed of the car and its position on the road. The control of speed is not unlike the example we discussed above, but this time the measurement of the output speed is given by the speedometer. The driver calculates the error, and actuation is accomplished by means of a pedal-operated throttle. So if the car is moving too slowly the throttle is opened, and if too fast the throttle is closed.

When controlling the position of the car, the driver knows that he must keep to his lane (Fig. 4.12). So the road itself is the input to the system, and the objective of the position control system is to make the car's position coincide with the road's position. The driver's eyes measure the position of the car and of the road, and his brain calculates the error. Corrective action is taken by means of the steering wheel. If the car is too far to the right, the driver steers left; if too far to the left, he steers right.

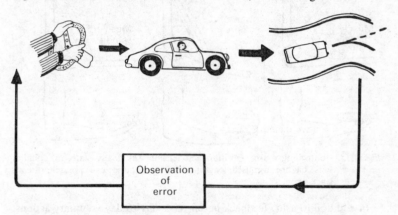

Fig. 4.12. Driving a car is a closed loop operation.

Besides being a useful component in a control loop system, the human operator is himself a nice example of a closed loop system. Consider the simple task of lifting a pencil from the table. This requires a control system to make the hand's position coincide with that of the pencil. The pencil's position is the input to the system and the hand's position is the output. Your eyes and brain measure these positions, determine the error and then order the muscles to take corrective action, i.e. to reduce the error to zero. So lifting a pencil is a closed loop operation.

The means of controlling muscle position and tension is a most intriguing one. A typical muscle contains tens of thousands of **fibres**, each about a tenth of a millimetre in diameter and a couple of centimetres in length. They are the motor elements that provide the pull. But in addition to these fibres there are hundreds of **spindles**, even thinner than the fibres. They act as measuring elements and are capable of detecting changes in length.

Fig. 4.13(*a*) shows how our muscles can maintain a fixed position in spite of an increasing load. When the load is increased, as in Fig. 4.13(*b*), the muscle stretches and the spindles sense the change in length. The spindle then sends nerve impulses to the spinal cord where they impinge on a motor cell at a synapse. This excites the synapse to send a motor impulse back down to the muscle, causing it to contract again (Fig. 4.13(*c*)).

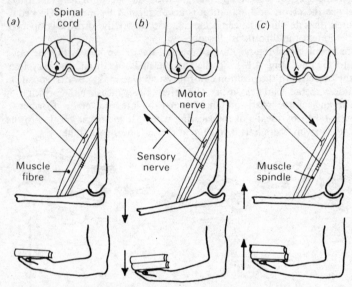

Fig. 4.13. Controlling muscle position and tension: (*a*) steady state; (*b*) spindles sense change in muscle length; (*c*) motor impulses cause muscle to contract again.

In addition to using feedback in our basic day-to-day voluntary actions, the body has also many closed loop systems controlling its physiological processes. Take body temperature as an example. We are at our best when our body temperature is 37°C, and even though the surrounding temperature may change by as much as 60°C, our inbuilt control systems can maintain the desired 37°C. This process is known as **homeostasis**. The system uses negative feedback. It is error controlled and if the body is too hot or too cold, corrective action is taken. If too hot we perspire and heat is lost by evaporation. In addition, blood flow to the surface of the skin is increased and we flush. If it is too cold we shiver, and the heat generated by the muscles helps to increase the body temperature. This is helped by a reduction in blood flow to the surface of the skin.

So the body employs many different kinds of feedback mechanisms, but in spite of this, as an element in a feedback system the human operator has several shortcomings. This can be demonstrated by referring to the task of tracking a moving target. It is well known that a marksman can hit clay pigeons consistently. This requires him to aim ahead of the pigeon to allow for its movement during the period of flight of the bullet. Compare this with shooting down a military aeroplane that can move a distance equal to its own length in less than one tenth of a second. This was the task faced by gunners during the Second World War. It is even more demanding nowadays, with aircraft speeds up to 1000 m/s. Man is not equal to this task and so it has become necessary to develop automatic equipment.

The limitations that preclude us from such tasks have been demonstrated many times using the simple apparatus shown in Fig. 4.14. A man is seated before an oscilloscope which shows two spots of light. One spot is the reference or input signal and it moves at random left and right across the screen. The other spot represents the output of the human operator. Its position is controlled by the man by means of a joy stick which is a movable vertical rod similar to that used by pilots. The operator's task is to keep the output spot on top of the input spot. It is a tracking task not unlike the one you may have seen in amusement parks when you are challenged to keep a model car on course on a winding road as it moves rapidly past you.

This spot (the output) controlled by operator

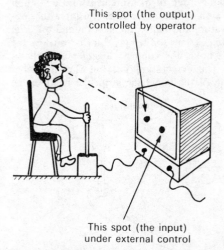

This spot (the input) under external control

Fig. 4.14. Closed loop with visual feedback; operator has to maintain spots in coincidence.

This experiment requires visual feedback and represents the sort of task that is faced by the anti-aircraft gunner, the driver and the pilot. The result of extensive tests of this nature have highlighted several human limitations. First there is a time delay of about 0.25 seconds in the human response. This is the time required for signals to travel from our sensors to our brain, to be processed and then to return to the actuating muscles. So, if you are driving a car at 100 km/h and you suddenly spot a hazard ahead, you will travel about 7 m before anything happens. This delay is even more serious for the modern fighter pilot who can cover 250 m before he can react. Fatigue, boredom, stress and intoxication can make matters even worse.

These tests have also shown that Man can track quite well when the target does not move too rapidly. But if the target position changes substantially in less than half of a second, then our performance begins to deteriorate. In other words, we will not be able to follow input signals that include frequencies greater than 1 Hz.

4.8   Automatic Control

These limitations on our control abilities, and the earlier mentioned limitations on our force-generating ability, have eliminated the human operator from many control systems. When there is no human in the control loop the system comes under automatic control. Let us look at some examples of automatic control systems.

We discussed earlier the problems of maintaining a constant temperature in a greenhouse. A human operator could do this by continually measuring the temperature with a thermometer and turning the heat on and off accordingly; but we can take the burden of measurement and actuation from the human by making the system an automatic one, using a thermostat both to measure the temperature and switch the heat on and off. Fig. 4.15 shows its mode of operation. When the temperature is too low the bimetallic strip (see Chapter 3) holds the contacts together and allows power to flow to the electric fire. The temperature rises, but when it reaches a level determined by the design of the bimetallic strip, the bend of the

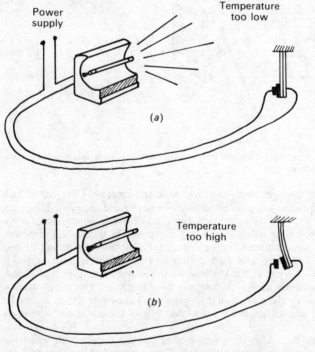

Fig. 4.15. On-off automatic temperature control: (*a*) too cold; (*b*) too hot.

strip causes contact to be broken, power is removed from the fire and the temperature begins to fall. When it has fallen by an amount, again determined by the design of the strip, the strip will straighten and close contacts again. You should be able to identify this as a negative feedback system. It seeks to reduce the error between a desired temperature (the **input** set by the design of the strip) and a measured temperature (the **output**, the temperature of the greenhouse). The figure emphasises its closed loop nature.

Let us now move from the control of temperature to the control of speed. During the eighteenth century the increasing use of wind power and steam power led engineers to search for means of controlling the speeds of their engines. How could they maintain the output speed at a constant level in the face of changing loads and other **disturbances**? As we have seen in Fig. 4.10, this could be done by a tireless human operator, but engineers wanted a more reliable means. The **centrifugal governor** was the answer. Its principle of operation is illustrated in Fig. 4.16. The axis of the governor is driven round by a belt connected to the drive shaft of the engine. Two masses M rotate with the governor, and it is these masses that measure the output speed and take corrective action. The faster the engine goes, the faster the governor turns, and, because of centrifugal force, the further the masses move away from the axis of rotation. Fig. 4.16 shows how this effect is used to correct the engine speed. The masses are connected by a linkage to a collar that is free to move up and down on the governor's axis. The collar, in turn, is connected through another linkage to the steam valve. When the speed is too high the masses move outwards and upwards; the collar is pulled downwards; this makes end A of the link to the valve move upwards; this rotates the valve plate clockwise thereby reducing the amount of steam going to the engine; the engine slows down. When the engine speed is too low this action is reversed.

So here again we have a negative feedback system that attempts to reduce the error in speed to zero. Its closed loop nature can be seen from the

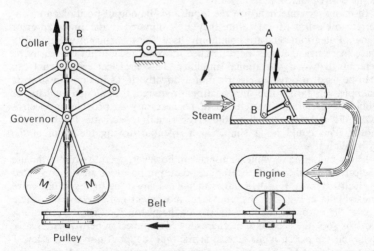

Fig. 4.16. Watt's governor in an automatic speed control system.

figure. The desired speed, which is the input to the system, could be varied by adjusting the lengths of the links AB.

The invention of the centrifugal governor is often attributed to James Watt. The one shown in Fig. 4.16 was designed by him in 1788, and the system of linkages in the figure are taken from a design sketch by Boulton and Watt in 1798. There is evidence, however, that Watt was inspired by Thomas Mead's earlier work on the centrifugal pendulum. Watt himself admitted that the application of the centrifugal principle was not a new invention and had been applied by others to the regulation of water, windmills and other things. But he had greatly improved the mechanism by which it acted on his engines.

Servomechanisms

The servomechanism is another common form of automatic control. Servomechanisms (from the Latin word for slave) are automatic feedback control systems in which the motion of an output member is constrained to follow the motion of an input member, with power amplification. An example is the system that controls the position of the rudder of a large ship. (It takes us right back to the origins of the word 'cybernetics'.) Here the input is the position of the helmsman's wheel, and the output is the position of the rudder. Ideally the rudder should follow the wheel, and it is obvious that an enormous degree of power amplification is needed.

Other examples are: (*a*) control of a radio telescope, where the output is the direction in which the telescope is pointing, and the input is the position of the target star; (*b*) control of an aircraft control surface where the output is the position of the control surface, and the input is the position of the pilot's control column; (*c*) control of an automatic lathe, where the output is the position of the cutting tool with respect to the workpiece, and the input or desired position can come from a computer tape or a template cut to the desired shape.

In many servomechanisms, the input and the output positions are converted to electrical form before they are compared to determine an error signal. Potentiometers are commonly used for this purpose (Fig. 4.17). Once in electrical form, the difference in input and output can easily be determined by an operational amplifier. This electrical error signal can then be used to drive an electric motor directly. If, as in the ship steering example, very large force amplification is required, then an hydraulic motor would be preferable, and it would be necessary to provide an electro-hydraulic interface to allow the error signal to activate the hydraulic motor. This could be as simple as a solenoid driving the spool of the hydraulic valve.

Many hydraulic servomechanisms can, however, operate without the use of electrohydraulic interfaces. Input and output positions are not converted to electrical form. Fig. 4.18 shows such a system, using the hydraulic valve and cylinder of Fig. 4.8. The input is the movement x of the valve's movable sleeve; the output is the movement y of the piston rod. You have to look hard to spot the feedback mechanism. The connector between the valve stem and the piston is the clue. Imagine that the input moves the sleeve 1 cm to the right. This connects the oil supply to the left-hand end of the

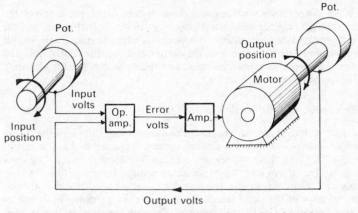

Fig. 4.17. An electromechanical servomechanism for control of position.

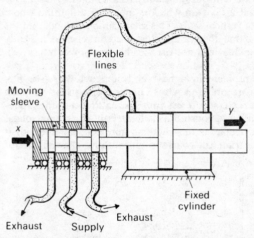

Fig. 4.18. An electrohydraulic servomechanism. The output follows the movements of the valve sleeve.

fixed cylinder, and simultaneously connects the right-hand side to the exhaust side. The pressure on the left-hand side of the piston is therefore high, and that on the right-hand side is low. The moving piston is therefore pushed to the right, the output y being in the same direction as the input x. But the beauty of this arrangement is that it not only moves y in the same direction as x, but it makes y equal to x. You can see this from the figure. As the piston moves to the right it carries the valve stem with it, and this gradually closes the valve ports. You will recall that the input x initially moved the valve sleeve to the right by 1 cm. So when the output y has moved to the right by 1 cm it will make the valve stem catch up with the movement of its sleeve. The valve ports will then be closed, no oil will flow to the cylinder and its piston will therefore stop at a displacement of 1 cm. So output equals input.

Such a system finds wide application. In a copy lathe, for example, the input x would be generated by a template cut to the desired shape, and the output y would be the cutting tool position. Note that the valve fulfils three functions: its stem measures the output position, the relative position of stem and sleeve measures the error, and the valve supplies the actuating power to the cylinder.

Early Examples of Automatic Control

You had to look hard to recognise the closed loop nature of the above system. Some of the earliest feedback systems were just as hard to recognise. Let us go back to the second century BC when the great Ctesibius was alive. He was one of the Alexandrian school, the output of the University of Alexandria, west of the Nile delta. Archimedes, Euclid and Hero were old boys of the same school. The water clock or **clepsydra** is one of the best known of Ctesibius' inventions. Its principle of operation is illustrated in Fig. 4.19(*a*). Water trickles in steadily to a container, and as the level rises it carries with it a float on which is mounted a figurine whose hand points the hour on a vertical scale. This seems simple enough, but the difficulty lies in ensuring that the water enters the container at a steady rate. If it doesn't then the float will not rise steadily and some 'hours' will be longer than others.

Ctesibius solved this problem in a neat and compact way (see Fig. 4.19(*b*)). His solution was recorded in AD 37 by Vitruvius in his book *De Architectura*. It gives you a chance to test your Latin: 'Praeclusiones aquarum ad temperandum ita sunta constitutae. Metae fiunt duae, una solida, una cava, ex torno ita perfectae, ut alia in aliam inire convenireque possit et eadem regula laxatio earum aut coartatio efficiat aut vehementem aut

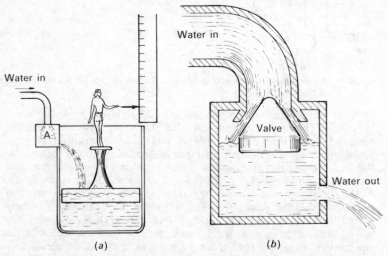

Fig. 4.19. (*a*) The water clock or clepsydra of Ctesibius. (b) The float maintains a constant water level.

linem in ea vasa aquae influentem cursum.' How did you do? 'Valves to regulate the flow of water are constructed thus: two cones are made, one solid, one hollow, finished on a lathe in such a way that one can enter and fit into the other and that by the same principle, their loosening or tightening will produce a strong or a weak current of water flowing into the vessel.'

The flow of water from the orifice in Fig. 4.19(*b*) will be constant only if the depth of water above the orifice remains constant. So Ctebisius realised that the solution rested in maintaining a constant level of water in his regulator. The float valve was a nice solution. If the level is too high the float moves upwards, restricting the flow into the regulator vessel and so allowing the level to fall. The reverse happens if the level is too low. So this is a feedback control system that attempts to maintain the level at a set value. The input or desired level is determined by the position of the fixed cone. The valve float measures the output position (the level). Its position, relative to the fixed cone, determines the error. And in addition to these two functions the valve also provides the actuating means, the flow rate, to correct the error. Variants of Ctebisius' system are used to this day in the carburettor float chamber and in the domestic water closet.

Another interesting example of an early feedback system is provided by the digester, or **pressure cooker** (Fig. 4.20). This was invented in 1681 by the French doctor, Denis Papin. (We mentioned his work on the vacuum in Chapter 1.) Papin realised that food cooks more quickly when the temperature is high. But, at atmospheric pressure, water boils at the relatively low temperature of 100°C. How could the temperature of boiling water be raised? You will now know from basic physics that the boiling point of

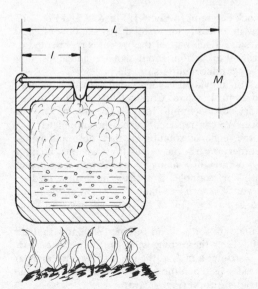

Fig. 4.20. Papin's pressure regulator.

water increases with pressure, and Papin used this fact in his digester. By enclosing the food and water and heating it, he allowed the pressure to rise to several atmospheres, thereby increasing the temperature of the boiling water. Now there were obvious dangers in such a procedure, for we all know that pressurised steam can be lethal, and exploding boilers have caused many terrible accidents. This is why Papin had to devise his pressure regulator, or safety valve.

Fig. 4.20 shows that the idea is quite a simple one. A lever is attached to the lid of the pot, and a weight at the end of this lever holds a plug in a hole in the lid. When the pressure gets too high the force on the plug is sufficient to lift it off its seat, thereby releasing steam and allowing the pressure to fall again. We can calculate the input or desired value of pressure as follows. If the area of the hole is a, then the force acting on the plug is pa. Using our knowledge of levers we can easily show that the plug will lift when the moment pal exceeds the moment MgL, or when $p > MgL/al$. So this is the input. The moment of the plug force is the measure of the controlled pressure; the difference between this and the moment of the weight is the measure of the error.

4.9   Quantitative Analysis of Control Systems

It is difficult to get across the full advantages of feedback control without recourse to calculation, and fortunately we can do this adequately without getting involved in high-powered mathematics. We will restrict ourselves to arithmetic, algebra and the concept of the **block diagram**. Nevertheless, for those readers who detest mathematics the escape route is to jump to Section 4.10.

We have already met block diagrams in Fig. 4.1, Fig. 4.11 and Fig. 4.17. They are intended to show in an easily understood way the flow of cause and effect throughout a system. Each part of the system is represented by a block, each block indicating an action. Input signals, the causes, flow into these blocks where they produce outputs or effects.

Let us look at examples of individual elements that we will later draw together to make a speed control system. Consider first a potentiometer that is used to convert rotary movement into voltage. The volume knob on your radio is such a device. We can represent such a potentiometer by a block whose input is θ, the angle of rotation of the knob, and whose output is a voltage V. In order to carry out calculations we must be more precise, and it is necessary to know how many volts are produced for each radian turn of the input. For example, a potentiometer described by $V = k_1\theta$ will produce k_1 volts for every radian (57.3°) turn at the input. This potentiometer is said to have a **gain** of k_1 volts/radian or k_1 V/rad (Fig. 4.21(*a*)).

Consider now a block diagram for a steam turbine. We know that the faster the steam goes in the faster the turbine will turn, so in this case we can take Q, the flow rate of steam, in m³/s as the input, and ω, the angular velocity of the turbine, in rad/s, as the output. We can write $\omega = k_5Q$, and the turbine has a gain k_5(rad/s)/(m³/s) (Fig. 4.21(*b*)).

It is not always possible to describe components by a block with a constant written inside it. In many cases the relationship between input and

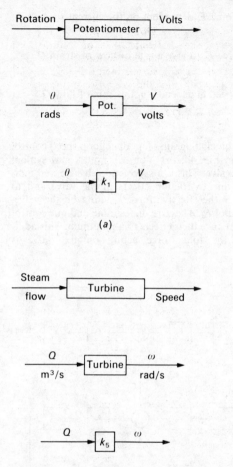

Fig. 4.21. Block diagrams of (*a*) a potentiometer and (*b*) a steam turbine.

output is not a simple proportional one, and dynamic effects and non-linearities will have to be accounted for. The complicated expressions inside such blocks are called **transfer functions**, but this however is beyond the scope of this book.

Let us attempt to put together a control system whose objective is to control the speed of a steam turbine by turning an input control knob. We shall use the following components:

(1) A potentiometer to convert θ, the rotation of the knob, to V, a voltage

$$V = k_1\theta$$

(2) An amplifier to convert this voltage V to current i

$$i = k_2 V$$

(3) A solenoid to produce a force F_s dependent on the current i

$$F_s = k_3 i$$

(4) A valve, operated by a force F, to give a rate of flow of steam Q

$$Q = k_4 F$$

(5) A turbine whose output speed ω depends on this rate of flow Q

$$\omega = k_5 Q$$

With these components the block diagram of an open loop speed control system would look like that in Fig. 4.22(*a*). You will note a new symbol between the solenoid and the valve. This is a **subtraction element** and indicates that disturbing forces, due to friction effects in the valve and to pressure variations, can reduce the effective force F acting on the valve. These are signified by the symbol F_d. They are undesirable and our control system should be designed to reduce their effects to a minimum. This additional element shows that the total force acting on the valve is $F = F_s - F_d$.

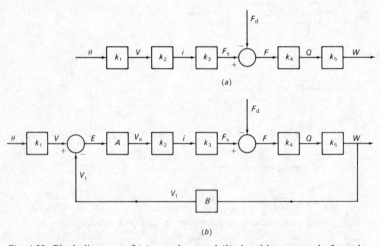

Fig. 4.22. Block diagrams of (*a*) open loop and (*b*) closed loop control of speed.

We can now do some calculations; working back from the output we have

	$\omega = k_5 Q$
or	$\omega = k_5(k_4 F)$
or	$\omega = k_5 k_4(F_s - F_d)$
or	$\omega = k_5 k_4 F_s - k_5 k_4 F_d$
or	$\omega = k_5 k_4(k_3 i) - k_5 k_4 F_d$
or	$\omega = k_5 k_4 k_3(k_2 V) - k_5 k_4 F_d$
or	$\omega = (k_5 k_4 k_3 k_2 k_1)\theta - (k_5 k_4)F_d$

Now to help us to continue, let us give values to these constants. Take, for example

$$k_1 = 0.1 \text{ V/rad}$$
$$k_2 = 5 \text{ A/V}$$
$$k_3 = 5 \text{ N/A}$$
$$k_4 = 5 \text{ m}^3/\text{N}$$
$$k_5 = 10(\text{rad/s})/(\text{m}^3/\text{s})$$

Our open loop speed control system is thus described by the equation

$$\omega = 125\theta - 50F_d$$

So if there are no disturbances ($F_d = 0$), a one radian turn of the input knob would cause the output speed to change by 125 rad/s, or 1194 rev/min (since one revolution equals 2π radians or $180°$). The possibility of disturbances, however, presents a problem, for an increase of friction in the valve of 1 N (i.e. $F_d = 1$ N) would reduce the speed of the turbine by 50 rad/s (477 rev/min). This confirms our earlier discussion of open loop control systems which emphasised that susceptibility to disturbance was one of their major disadvantages.

Can we reduce the effect of these disturbances by closing the loop? Let us design a closed loop system that still maintains the desired relationship between input and output, $\omega = 125\theta$, but reduces the effects of disturbances by a factor of 10. We are thus aiming for a system described by the equation

$$\omega = 125\theta - 5F_d$$

In order to close the loop it is necessary to measure the output speed ω, and we will assume that is done by a tachogenerator that provides a voltage proportional to speed. The tachogenerator equation is

$$V_t = B\omega$$

The closed loop block diagram is shown in Fig. 4.22(*b*). This shows that the voltage V_t is subtracted from the voltage V generated by the input potentiometer. Remember that negative feedback requires this subtraction. The difference between these signals is called E, the actuating signal, where $E = V - V_t$. An additional voltage amplifier has also been included, with the equation $V_e = AE$.

Our task is to determine the values of A and B required to satisfy the relationship

$$\omega = 125\theta - 5 F_d$$

Working backwards from the output again gives

$$\omega = k_5 Q$$
$$\omega = k_5 k_4 F$$
$$\omega = k_5 k_4 (F_s - F_d)$$
$$\omega = k_5 k_4 k_3 i - k_5 k_4 F_d$$
$$\omega = k_5 k_4 k_3 k_2 V_e - k_5 k_4 F_d$$
$$\omega = k_5 k_4 k_3 k_2 AE - k_5 k_4 F_d$$

Now $$E = V - V_t$$

or $$E = k_1\theta - B\omega$$

Hence $\omega = k_5 k_4 k_3 k_2 A(k_1\theta - B\omega) - k_5 k_4 F_d$

or $\omega = k_5 k_4 k_3 k_2 k_1 A\theta - k_5 k_4 k_3 k_2 AB\omega - k_5 k_4 F_d$

or $\omega(1 + k_5 k_4 k_3 k_2 AB) = k_5 k_4 k_3 k_2 k_1 A\theta - k_5 k_4 F_d$

Hence $$\omega = \left[\frac{k_5 k_4 k_3 k_2 k_1 A}{1 + k_5 k_4 k_3 k_2 AB}\right]\theta - \left[\frac{k_5 k_4}{1 + k_5 k_4 k_3 k_2 AB}\right]F_d$$

This is a horrible equation, but it looks a lot better when we substitute our earlier values for k_1 to k_5. It becomes

$$\omega = \left[\frac{125A}{1 + 1250AB}\right]\theta - \left[\frac{50}{1 + 1250AB}\right]F_d$$

Now, we wish to have a system described by

$$\omega = 1250\theta - 5F_d$$

Hence equating terms we get two equations

$$\frac{125A}{1 + 1250AB} = 125 \quad \text{and} \quad \frac{50}{1 + 1250AB} = 5$$

These can be solved to give:

$$A = 10 \text{ V/V and } B = 9/12\,500 \text{ V/(rad/s)}$$

We now have a system that reduces the effects of disturbances by a factor of 10.

Another advantage of the closed loop system is its ability to reduce the effects of variation in the component parts of the system. This example can also illustrate this point. Ignoring the effects of disturbances ($F_d = 0$), we have

$$\omega = \left[\frac{k_5 k_4 k_3 k_2 k_1 A}{1 + k_5 k_4 k_3 k_2 AB}\right]\theta$$

Let us examine the effect of changes in the valve gain k_4. Substituting the earlier values for the other gains, we have

$$\omega = \left[\frac{250\,k_4}{1 + \dfrac{22\,500}{12\,500}k_4}\right]\theta$$

For $k_4 = 5$, this gives $\omega = 1250\theta$, as expected. But if blockage in the valve were to reduce k_4 to 2.5 then the relationship becomes $\omega = 113.6\theta$. So a 50 per cent reduction in the valve gain only causes a 9 per cent reduction in output speed. This is clearly an advantage of closing the loop, for if the loop were open as in Fig. 4.21(a), a 50 per cent reduction in the valve gain would have caused a 50 per cent reduction in output speed.

4.10   Modes of Control

We saw earlier that amplifiers can operate in an on-off mode or in a continuous mode. An example of the former was the switch which could only be on or off. An example of the latter was the capstan amplifier whose output force was proportional to the input force. Closed loop control systems are designated on-off or continuous, depending on which types of components are included in the loop.

Take as an example the use of a thermostat to control the temperature of a greenhouse. The thermostat is an on-off device; when the temperature exceeds a certain value it breaks the electrical circuit and switches off the heat; when the temperature is too low it makes the contact and switches on the heat (see Fig. 4.15). Such a system is simple and cheap and is thus of considerable attraction, but it has a major disadvantage: the output tends to oscillate. This can be explained with the help of Fig. 4.23(*a*). A room temperature T_d is desired. The thermostat is so designed that it will break contact at T_u, the upper limit, and will make contact when the temperature falls to a lower limit T_l. The figure illustrates the oscillation in the room temperature. When the heat is on, the room heats up until T_u is reached. Heat is then switched off and the room cools until T_l when the heat is then switched on again, and so on. The temperature oscillates between T_l and T_u. If the designer wishes to achieve greater accuracy by using a more sensitive thermostat which reduces T_u and increases T_l, then Fig. 4.23(*b*)

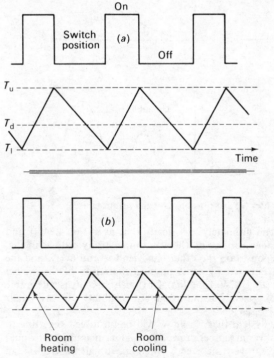

Fig. 4.23. Variations of temperature and switch position in a room temperature control system: (*a*) coarse control; (*b*) fine control.

shows that the frequency of switching increases. This can cause the thermostat and the heater to wear out more quickly.

So although on-off control can be cheaper than continuous control, there are problems of inaccuracy and wear to be considered. Because of this, systems requiring high accuracy, such as machine tool positioning systems, demand the use of continuous control such as that of Fig. 4.17 and Fig. 4.18.

4.11 Stability

Some of the advantages of closed loop systems have been identified above, but there is also one major disadvantage that has to be taken into consideration—that of the possibility of **instability**. When a system is unstable, its output runs away and control is lost. Some examples of stable and unstable responses are shown in Fig. 4.24.

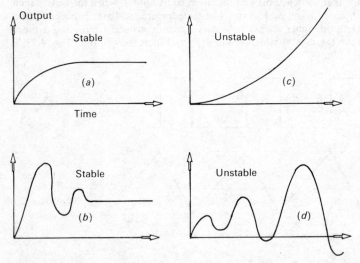

Fig. 4.24. System responses (*a*) and (*b*) stable; (*c*) and (*d*) unstable.

Stable responses tend ultimately to a steady state as in Fig. 4.24(*a*); and there may even be a few wobbles on the way to the steady state as in Fig. 4.24(*b*). Unstable responses on the other hand tend to run away, and the output grows until restricted by some physical limitation of the system such as motor power (Fig. 4.24(*c*) and (*d*)). Obviously stability is a desirable property of a control system but unfortunately instability can easily rear its ugly head. What causes it?

Let us construct a system that we know will be unstable. Reverting to our earlier example of room temperature control let us imagine what would happen if the thermostat were incorrectly wired up, so that the electric fire were switched off when the temperature was too low, and switched on when too hot. This is clearly a crazy system, and we would expect the fire either to stay on indefinitely or stay off indefinitely. The temperature of the room would run off to one or other of two extremes. Control would be lost and the system would be unstable.

Now if we examine this silly system closely we should be able to spot the root of the problem. Positive, rather than negative feedback, is being used! The output is being added to, rather than substracted from, the input. It is similar to the positive feedback loudspeaker system referred to earlier, except that whereas the temperature would steadily increase or decrease to a limit in this example, the sound output from the microphone would run away to an uncontrolled oscillatory howl. (We shall meet other examples of positive feedback in Chapter 6.)

You may justifiably comment, of course, that it would be a daft designer who would pursue such ridiculous schemes, with their obvious positive feedbacks. But Nature is devious and positive feedback can pop up in the most unexpected places. And in order to keep you on your guard it is necessary to have a brief look at the dynamic performance of the elements within a control system.

Effect of Lags

In our discussions of block diagrams we assumed that an input to a block would give the required output immediately. This is not the case in practice, for Nature always requires some time to respond. If we immerse a thermometer in a hot liquid, the mercury does not immediately jump to its final reading; time is needed for the heat to pass through the glass to the mercury, and for the mercury to expand. Similarly, when you depress the throttle in your car, its speed does not instantaneously rise to a new level: time is required for the additional fuel to get to the engine, and for the additional power to accelerate the mass of the engine and of the car. So our earlier assumptions about block diagrams have to be viewed with some suspicion. For example, in Fig. 4.22, the steam valve was shown as a block with an input F, a force, and an output Q, a flow rate. This was described by the equation $Q = 5F$, and you can see that this implies that, for example, as soon as F jumps from 1 to 2 N, then Q will jump from 5 to 10 m³/s. Such an instantaneous rise in flow would not occur in practice; there would be a **delay**, or **lag**, between input and output. Such lags are at the heart of the instability problem. If there are enough of them, they can turn the desired negative feedback into unwanted positive feedback.

Many systems, such as loudspeakers, break into violent oscillation when they become unstable, so some idea of the response of components to oscillatory inputs is necessary if we are to understand the phenomenon of instability. Fig. 4.25 will help us. It shows a component being excited by an oscillating input signal. If there were no lag in the response the output would be in step with the input, and would look like that in (*a*). But if the dynamic properties of the component cause a considerable lag, then the output could look like (*b*) or (*c*). In (*b*) the output lags the input by one quarter of a period, and it can be seen that the output is zero when the input is at its extreme values. In (*c*) the output lags by a half period; it is at a maximum when the input is at its minimum, and vice versa. So when the

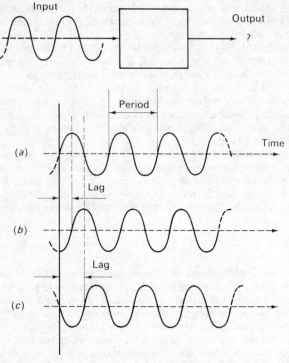

Fig. 4.25. Lags in a dynamic system: (*a*) no lag; (*b*) output lags input by a quarter
period; (*c*) output lags input by one half period.

lag is one half period the output is the negative of the input. You should
now begin to see the plot unfolding—another negative sign could turn
negative feedback into positive feedback! Read on.

These lags, or phase shifts, vary with the frequency of the input oscillation. In general the lag will increase as the frequency increases, because the
system finds it more and more difficult to keep up with the fast changing
input. So, for example, a component's output may lag its input by a quarter
of a period at 2 Hz, and by half a period at 5 Hz. The actual size or
magnitude of the output signal varies with frequency too, usually getting
smaller as the frequency increases. Sometimes, however, there are frequencies that the component 'likes'. These are called **resonant** frequencies, and
the outputs at these frequencies can be quite large.

Instability

We now have sufficient information to allow us to explain instability in a
closed loop system. The possibility of instability in such a system will be
proven by demonstrating that an oscillatory output can sustain itself without need of an input to the system (Fig. 4.26). First we have to assume
that an oscillating signal E exists at position 1. We then assume that the

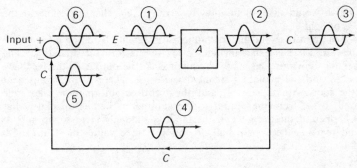

Fig. 4.26. A sine wave can sustain itself if element *A* has a gain of unity and introduces a phase lag of one half period.

element *A* introduces a lag of one half of a period at that frequency, and
that the size of the signal is unchanged, i.e. the gain is unity. So the output
signal *C*, looks as shown at 2: it is in fact the negative of *E*. This signal
passes out of the system at 3, but it is also fed back through 4 and 5 to the
subtraction element. There is no input to the system so the output *E* of the
subtractor will be the negative of *C*. Surprise! Surprise! The signal at 6 is
the same as the one we started with at 1. So we conclude that if the component has a gain of unity and causes a lag of one half of a period, then
the closed loop system can sustain an oscillation indefinitely. Such a system
would be little use as a control system. The problem basically is that the
lags in the system have changed it from a negative feedback to a positive
feedback system. Two minuses, one at *A* and one at the subtractor, multiply
to give a plus, and we have a system that behaves like the loudspeaker we
mentioned earlier.

As shown, the system is **critically stable**. If the gain of *A* were less than
1, the system would be stable and the oscillation would gradually die away.
If, however, the gain were greater than 1, the magnitude of the signals
would grow each time they moved around the loop; the response would
run away and the system would be unstable.

So the stability of closed loop systems is affected by the gains of the
individual components and by the lags that they introduce. The designer
has to watch this carefully. He may wish to make gains as high as possible
in order to produce a sensitive system with a fast response. But he cannot
go too far. He may wish to add extra components to the loop, but he must
be careful not to make the lags too high.

Effect of Dead Time

There is a particular form of time-lag, called **'dead time'**, during which
nothing at all happens at the output. We have seen earlier that there is a
dead time of about 0.25 seconds in control systems that employ human
operators, for humans have a reaction time of about 0.25 seconds. This is
the time for information to be transmitted around the nervous system.
Dead times increase when information transmission distances become
large. For example, a message from a robot on Mars to a human controller

on Earth would take 20 minutes to arrive, even when travelling at the speed of light.

We can reinforce our understanding of instability by doing a few calculations for a system that includes a lag in the form of dead time. The system is shown in Fig. 4.27. The input is R, the output C, the feedback signal B and the actuating signal $E = R - B$. There is one component in the loop with a gain A, and the feedback information takes one second to get from the output to the subtractor. Let us calculate what will happen at the output C if the input R suddenly increases from 0 to 1 and remains there. We shall do this for three values of A: 0.75, 1.0 and 1.1.

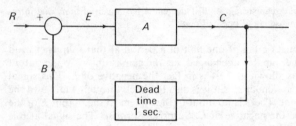

Fig. 4.27. A simple system with a dead time in the feedback path.

First $A = 0.75$:

Time (s)	R	B	E	C
0	1	0	1	0.75
1	1	0.75	0.25	0.188
2	1	0.188	0.812	0.609
3	1	0.609	0.391	0.293
4	1	0.293	0.707	0.531
5	1	0.531	0.469	0.352
6	1	0.352	0.648	0.486

The table is constructed as follows. For the first second, i.e. for time 0 to 1, the input R is 1 and the feedback signal B is 0, since B cannot respond until at least 1 second has elapsed. If B is 0 then $E = (R - B)$ will be 1. If E is 1, then $C = (0.75E)$, will be 0.75. This completes the first row of the table. The second row covers the time interval 1–2 seconds. Again $R = 1$ since the output is fixed. The feedback B over the interval will be equal to the output C over the interval 0–1 seconds, since there is a dead time of 1 second. Hence the value of B in the second row equals the value of C in the first row. The actuating signal E is then $R - B = 1 - 0.75 = 0.25$, and the output $C = 0.75E = 0.188$. And so on. The response is plotted in Fig. 4.28(a). It is a stable response with the output ultimately settling out to a steady value.

We can construct a similar table to show that, if the gain is increased to unity, the system is critically stable with a response indicated in Fig. 4.28(b). Further increase in gain to 1.1 produces an unstable response (Fig. 4.28(c)). This confirms our earlier deduction that increasing the gain of a system that includes lags will untimately lead to instability.

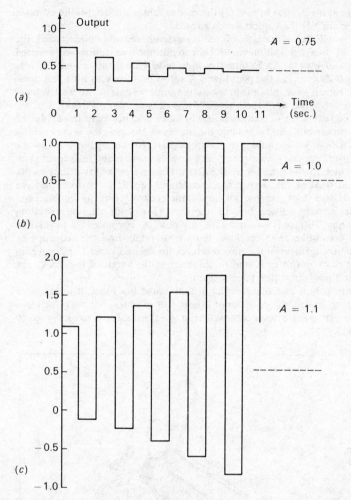

Fig. 4.28. Response of a system with dead time for various values of gain: (*a*) 0.75, stable; (*b*) 1.0, critically stable; (*c*) 1.1, unstable.

4.12 Controlling the Order of Events

So far we have been concerned with methods of controlling a single variable such as temperature or velocity. There is another important area of control—that concerned with controlling the order and the timing of a **sequence of events**. Many manufacturing processes require this form of control. For example, an automated drilling operation may include the following sequence of events: (1) the component is pushed into position, (2) the component is clamped firmly, (3) the component is drilled, (4) the clamp is released, and (5) the component is ejected.

The wonderful automata of the nineteenth century, and indeed the robots of this age, all have to be programmed to follow an ordered sequence of events. An automaton may have had to lift a pen, turn its head, roll its eyes, set the pen down. A robot may have to lift a hot component, turn it over, place it in a stamping machine. How do we tell these machines what we want them to do? How do we program them?

The earliest form of automata used clockwork, not only to supply the motive power but also to provide the means of ensuring the proper timing of each action. What could be more appropriate for timing but a clockwork mechanism? An early example of such a mechanism is the tea's maid (Fig. 4.29), which at the time set by the alarm boils water, pours it into a cup, and then wakens the sleeper for a welcome cup. It is obvious that the hands of the clock are the all important control elements in this case. Consider also the musical box. Its operation (Fig. 4.30) relies on a rotating drum whose surface is covered with tiny pins. As it rotates the pins strike at tiny comb-like teeth, each with its own pitch, and the sequence or ordered arrangement of the pins produces the desired tune. The drum is in fact the clock in this case, and the pins act as the hands of the clock. The drum and pins constitute the program.

Now in robots and other forms of automated machines, it is necessary to produce an orderly sequence of events, not of notes, but the principle is the same. In many cases a camshaft (Fig. 4.31) acting as a clock is used to

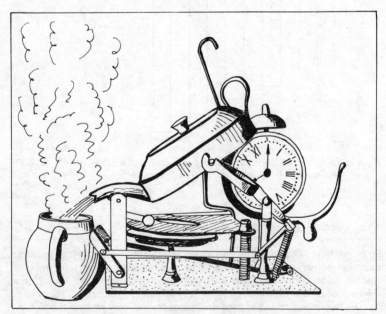

Fig. 4.29. A tea's maid—an early example of sequence control.

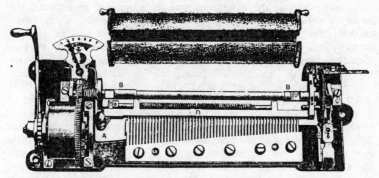

Fig. 4.30. A musical box—an early example of synchronous control.

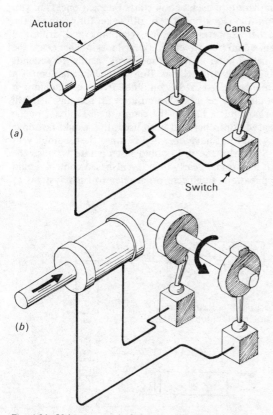

Fig. 4.31. Using a cam shaft for synchronous control.

control the movements of actuators. You will note the similarity to the mechanism of the musical box, but this time the pins (the cam lobes) are used to initiate the extensions and retractions of an actuator that could be electric, pneumatic or hydraulic. The **program**, or the timing of the events, can be altered by adjusting the relative angular positions of the cams. Other programmers use punched tape or cards to store the instructions, and these are moved past a reading head that detects the presence or absence of holes, and initiates the appropriate action.

All of the above sequence controllers have a clock, in one guise or another, at the heart of operations. Because of this they are known as **synchronous controllers**. Another class, the **asynchronous controllers**, work differently. They use the completion of one event as the trigger to start the next event, and for this reason they are also often referred to as **event-based controllers**. Fig. 4.32 shows a three-actuator sequence, operating under asychronous control. The operator presses button *a* momentarily; this activates actuator A which extends to press button *b* and returns; that activates actuator B and so on. Finally button *d* could be used to start the cycle again. So the completion of each action starts the next one. Can you see how this differs from the clock-based control? Take, for example, the sequence described earlier concerning feeding, clamping and drilling a component. The sequence could be driven by a camshaft, in which case the various events would be initiated at fixed times—say, clamp 10 seconds after feed, and drill 10 seconds after clamp. But what would happen if a component jammed while being fed in? It would not arrive in the clamp in time. The camshaft would turn relentlessly and after 20 seconds the drill would descend and perhaps put a hole in the clamp or in the base of the machine. If an event-based controller had been used, this would not have happened. The clamp would not have operated because the feeding event had not taken place. Similarly, the drill would not operate because the clamping event had not occurred. Clearly then, event-based control would be superior in that application. Engineers often refer to such systems as being **interlocked**.

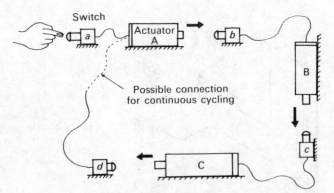

Fig. 4.32. Asynchronous or event-based control of three actuators.

4.13 Exercises

Numerical answers are given in Appendix 1.

1. Describe an on-off amplifier and a continuous amplifier, both using a fluid as the control medium. Give examples of their use.

2. Control systems use many different types of components—mechanical, electrical, electronic, hydraulic, pneumatic, fluidic. (*a*) List two advantages and two disadvantages of each type of component. (*b*) Give one example of each type.

3. (*a*) Explain the differences between open loop and closed loop control. (*b*) Describe two examples of closed loop control taken from your own body; identify the components and variables, and explain the negative feedback process. (*c*) Give two examples of positive feedback.

4. Compare the advantages and disadvantages of on-off and continuous control systems. Give two examples of each mode of control.

5. Propose automatic control systems for each of the following and draw block diagrams to explain their method of operation: (*a*) level of powder in a hopper; (*b*) flow of air to a furnace; (*c*) force applied by a testing machine.

6. Explain how the error is detected in: (i) Papin's digester, (ii) Watt's governor, (iii) Ctebisius' level controller, (iv) a domestic central heating system, (v) the cistern for a water closet.

7. (*a*) Choose one example of a control system and draw a block diagram to show how the system operates. (*b*) In the control system shown in the Fig. 4.33(*a*) the component gains are $A = 5$, $G = 2$, $H = 2$, $D = 4$. If the input $R = 1$ and the disturbance $U = 3$, determine the output C.

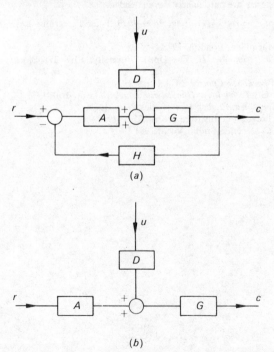

(*a*)

(*b*)

Fig. 4.33. (*a*) Closed loop system; (*b*) open loop system (for Exercises 7 and 8).

8. (*a*) In the open loop control system of Fig. 4.33(*b*) the component gains are $A = 5, D = 4, G = 2$. Find an expression for the output C in terms of the input R and the disturbance U. (*b*) If negative feedback is used as in Fig. 4.33(*a*) and the values of D and G are maintained, find the values of A and H so that the output bears the same relationship to the input as in part (*a*), but the effect of disturbances is reduced by a factor of 10.

9. (*a*) Explain how lags can arise in a control system. (*b*) Why is a lag of 180° equivalent to multiplication by -1? (*c*) Explain how increasing the sensitivity of a controller can cause instability.

10. (*a*) What is dead time? Give examples in machines and in humans. (*b*) What would be the effect of increasing the dead time in Fig. 4.27 from 1 to 2 seconds? (*c*) What would happen if the dead time were infinitely long? (*d*) If the dead time were zero and the input unity, what value would the output assume for $A = 0.75$, 1.0 and 1.1 (see Fig. 4.28)?

11. (*a*) Compare synchronous and asynchronous control of events. (*b*) a pneumatic hacksaw machine is to be driven by a reciprocating pneumatic cylinder; show how synchronous control can be used for this purpose. (*c*) In an electro-pneumatic system three pneumatic cylinders are operated by electro-pneumatic valves. Show how asynchronous control can be used to push a component into position under a drill, to clamp it, to drill it, to withdraw it.

4.14 Further Reading

Those marked with an asterisk are particularly recommended.

Bennett, S., *A History of Control Engineering, 1800–1930*, Peter Peregrinus Ltd, 1979.
Burke, J., *Connections*, Macmillan, London, 1978.
* *Fundamental concepts in technology II*, The Open University, PET 271 Block 2(5,6).
Mayr, O., *The Origins of Feedback Control*, MIT Press, 1970.
Pill, E. J., and Truxal, J. G., *Technology: Handle with Care*, McGraw-Hill, 1975.
Porter, A., *Cybernetics Simplified*, English Universities Press, 1969.
* *The Man-made World*, Polytechnic Institute of Brooklyn, 1968.
Trask, M., *The Story of Cybernetics*, Studio Vista, 1971.

5
Technology Extends Man's Brain

5.1 Introduction

The brain of the average young adult male has a mass of 1.41 kg whilst that of the average female is 1.03 kg. The largest brain ever recorded was that of the Russian author, Ivan Sergeyvich Turgenev (1818–83); its mass was 2.01 kg. But size isn't everything, for the dolphin's brain is bigger than ours, and yet it cannot talk!

We have already seen how our senses feed information about the outside world to the brain, and how the brain copes with that information. We have also seen how the brain is the central part of the human control mechanism. It is the cybernetic powerhouse. But there are three important functions of the brain that we have not yet mentioned: its ability to perform arithmetical calculations, its ability to make decisions, and its ability to memorise. The computer has extended all of these abilities.

5.2 Historical Survey

We do not know when Man first learned to count, but there can be no doubt that his fingers and toes were the first calculating aids. The modern words '**decimal**', related to the Latin word for ten, and '**digit**', related to the Latin word for finger, spring from those early origins.

Next to fingers and toes, the earliest calculating device was the **abacus**. The first of these was no more than a board dusted with a thin layer of dark sand on which one could trace numbers and figures. The Hebrew word for dust is *abaq*, and it seems likely that the Greek word *abax* (a flat board) was related. It is said that Archimedes was calculating on a sand board when he was killed by a Roman soldier in 212 BC.

Later, around the fourth century BC, a different type of abacus evolved in the Greek and Roman worlds. It was a board on which parallel lines or grooves marked the place values of a number system, usually decimal. Counters, in the form of pebbles, were moved back and forth on these lines or grooves. The Latin for pebble is *calculus*—hence our word calculate. The use of the abacus spread through the Arabic culture to India, the Far East and Russia. It reached China in the twelfth century, and it is still in use there today where it is called the *suan pan*. The Chinese symbol for *suan* (calculate—see Fig 5.1(*a*)) shows an abacus held below by the symbol for 'hands', and with the symbol for 'bamboo' above the abacus. The *suan pan* (Fig. 5.1(*b*)) has counter beads that move easily along bamboo rods. Each rod has five beads (ones) below the cross-bar and two beads (fives) above the bar. The operator indicates a given number by pushing the beads towards the cross-bar.

(a)

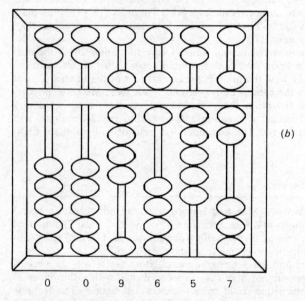

(b)

0   0   9   6   5   7

Fig. 5.1. (*a*) the Chinese symbol for 'calculate'. (*b*) The *suan pan* showing the number 009657.

The Japanese abacus, the *soroban*, differs slightly in that there is only one 'five' above the bar and four 'ones' below. Modern operators can achieve high speeds and although a long way short of the speed of the modern computer, they were able to compete with the earlier mechanical desk calculators. In 1946 a widely publicised match was arranged, and it was found that the Japanese abacist was faster in all calculations except the multiplication of huge numbers.

Chronologically the next major calculating aid was invented by John

Napier, the Scottish mathematician, at the beginning of the seventeenth century. It consisted of eleven graduated rods, known as **Napier's bones** (Fig. 5.2(*a*)). Multiplication could be accomplished by setting up a combination of the appropriate rods (Fig. 5.2(*b*)). This seems to us to be a very elementary device, but it is easy to underestimate its effect on commerce at that time. It allowed the ordinary tradesman to reduce a long multiplication to a simple addition, and that, in days of relative ignorance, must have been regarded as a big step forward.

But Napier was to make a much greater contribution to the development of calculating devices. He invented **logarithms**, and these too reduced the task of multiplication to the simpler one of addition. Napier's logarithms were invented in 1614 and were calculated by William Briggs to 14 decimal

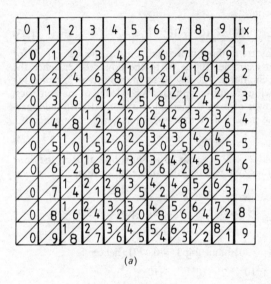

(*a*)

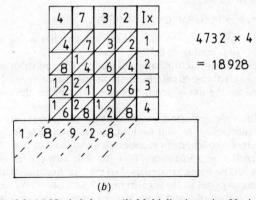

4732 × 4

= 18928

(*b*)

Fig. 5.2. (a) Napier's bones. *(b)* Multiplication using Napier's bones.

places in 1624. Logarithms can be explained as follows. Consider the fairly obvious arithmetical statement

$$(10 \times 10) \times (10 \times 10 \times 10) = (10 \times 10 \times 10 \times 10 \times 10) \quad (5.1)$$

This can be rewritten in the following mathematical shorthand

$$10^2 \times 10^3 = 10^5 \quad (5.2)$$

where 10^5, for example, means that five tens are multiplied together, and is often referred to as 10 raised to the **power** of 5, or 10 to the 5. The important thing to note is that the power on the right-hand side of the equation is equal to the sum of the two powers on the left-hand side. Therein lies the clue that showed Napier how to convert the multiplication process into a simple addition.

In general we can write

$$10^x \times 10^y = 10^{x+y} \quad (5.3)$$

and Napier's logarithms are nothing more than a list of the x's and y's for different numbers. For example, 100 is 10 to the power of 2, hence the logarithm of 100 is 2. Similarly, 1000 is 10 to the power of 3, hence log $1000 = 3$. And less obvious to those without a knowledge of mathematics, 2 is 10 raised to the power of 0.3010, giving log $2 = 0.3010$. So the logarithm of a number is the power of 10 that gives that number. In general:

$$n = 10^{\log n} \quad (5.4)$$

Now we all know that $2 \times 3 = 6$, but let us show how this can be done with the aid of logarithms. We will call the answer z, so that

$$2 \times 3 = z$$

We rewrite 2, 3 and z as powers of 10, using equation (5.4)

$$10^{\log 2} \times 10^{\log 3} = 10^{\log z}$$

The tables give log $2 = 0.3010$ and log $3 = 0.4771$; hence

$$10^{0.3010} \times 10^{0.4771} = 10^{\log z}$$

Hence, using equation (5.3), we can write

$$\log z = 0.3010 + 0.4771 = 0.7781$$

We then look up the tables again to find which number has a log of 0.7781. The answer is 6 (the antilog of 0.7781).

That calculation showed that multiplication of numbers is equivalent to the addition of their logarithms. Similarly, division can be achieved by subtracting logarithms. So the use of log tables greatly simplifies the task of calculation.

The **slide rule** simplifies it even further by providing a device that carries out the addition and subtraction as well as 'looking up' the log tables. It consists (see Fig. 5.3(*a*)) of two logarithmic scales, so arranged that one can slide relative to the other. Fig. 5.3(*b*) shows how the slide rule carries out the calculation 2 times 3. The scales are designed so that the linear distance between numbers is proportional to the logarithm of the numbers. We saw earlier that to find the product of 2 and 3, we start by adding log 2 and log

3. This is done by displacing the top scale by log 3 relative to the bottom one as shown in Fig. 5.3(*b*). The sum of log 3 and log 2 is found where the log 2 mark on the top scale intersects with the bottom scale—in this case at the log 6 mark.

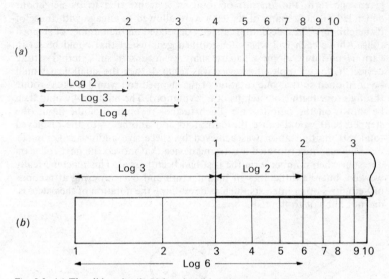

Fig. 5.3. (*a*) The slide rule. (*b*) Using the slide rule to calculate 2 × 3.

Although Napier invented logarithms, it was Oughtred who invented the slide rule, a circular one, in 1621. The more familiar linear version was produced in 1654 when it was in wide use for customs and excise calculations.

By its very nature, the slide rule is subject to error. It is an analogue device (see Chapters 3 and 6), representing numbers by length, and the accuracy of a calculation therefore depends on how accurately the lengths can be read. Errors can easily build up in long calculations, and although a degree of error may be acceptable in some cases, there are other cases when 4 times 6 must equal 24 and not 23.97 or 24.03. This led engineers to search for mechanical calculators that would be capable of precise mathematical results. It required a fresh look at digital machines, like the old abacus, that could deal with discrete numbers. If you add 2 pebbles and 3 pebbles you will always get 5 pebbles; never 4.97 pebbles.

5.3 Mechanical Calculators

The Frenchman, Pascal, invented his mechanical digital calculator in 1642. Known as the **Pascaline** it worked very well, and some examples still survive to this day. It used gear wheels and cams in a similar way to the odometer in a car's speedometer. Pascal's machine won him great renown throughout Europe, and this inspired many seventeenth-century mathematicians to attempt to produce better calculators. The most successful of these was the great mathematician Leibnitz who, in 1694, invented a calculating machine that used an extremely new device—the **stepped roll** (Fig. 5.4). The stepped roll was a spur gear wheel with nine teeth of different lengths so that, given one turn, the amount of rotation it transferred to its neighbour depended on how many of its teeth were allowed to engage with those of its neighbour. The calculating capacity depended on the number of stepped rolls. Above each roll was a ten-toothed gear wheel that could be set in various positions along a square shaft by means of an external setting device. If, for example, the device was set at 4, then the small gear would be positioned so that one rotation of the stepped roll would make it rotate through four teeth (four-tenths of a revolution). The number 4 would then be shown on the output scale. Each further revolution of the input (the stepped roll) would cause the output to add another 4 to the displayed total. So multiplication was achieved by successive additions. Leibnitz's machine also incorporated a carrying device that moved the next roller on one step when number 9 on the first had been passed. This machine really worked, but the difficulties of manufacture and overpowering attractions of Leibnitz's other interests, such as developing the notation of the calculus, did not allow it to flourish.

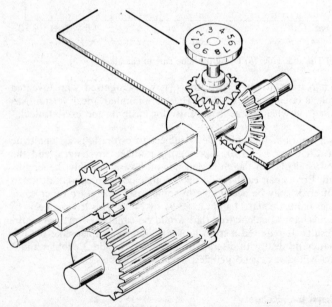

Fig. 5.4. The stepped roll was a feature of Leibnitz's digital calculator.

It was not until 1822 that the next real step in the development of computers arose. Charles Babbage (1791–1871) was a rich, talented and eccentric mathematician. He was quick to argue that routine arithmetical calculations ought to be relegated to the machine. He writes: 'One evening

I was sitting in the room of the Analytical Society at Cambridge . . . with a table of logarithms lying open before me. Another member coming into the room, and seeing me half asleep, called out, "Well Babbage, what are you dreaming about?" to which I replied, "I am thinking that all these tables might be calculated by machinery."'

Babbage's plans for his first machine, the **Difference Engine**, were published in 1822. It used the mathematical theory of differences to solve equations, but it too, like Leibnitz's machine, was beyond the manufacturing capabilities of the day. So in spite of financial support from the Government the project was stopped in 1833. (A version of this machine was built in 1843 by the Swede, Scheutz.) Babbage's major contribution, however, was his second machine, the **Analytical Engine**. This was his grand vision—a machine to perform any type of digital calculation—and the pursuit of this vision occupied the rest of his life. But it was never built, although a simplified model, made by his son, can be seen in the Science Museum in London.

Babbage's Analytical Engine was the true forerunner of today's computers. Fig. 5.5 shows the basic structure of a computer as we know it today (Babbage's machine had all of these elements):

(*a*) It used punched cards to get information into the machine (**the input**).

(*b*) It had an **arithmetic unit** that did the calculations (Babbage called this part the 'mill'); one addition could be carried out in 1 second.

(*c*) It had a **store** or memory where numbers could await their turn to be processed. This consisted of 50 counter wheels that could store 1000 numbers of 50 digits each.

(*d*) It had a **control unit** that organised the order of the calculations.

(*e*) Finally, it had an **output** device to display the results. This was in the form of either punched cards, direct printing or as a type set ready for print.

Babbage had planned to use punched cards for communicating with his

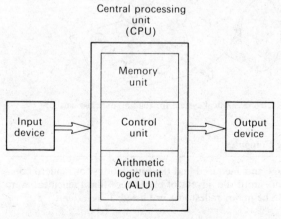

Fig. 5.5. The basic structure of a digital computer.

machine, but this idea had not been invented by him. It was the brainchild of Vaucanson, renowned for his automata, and was first successfully applied by Jacquard (1752–1834) who used it for producing intricate patterns on silk. His automatic loom was shown at the Industrial Exhibition in Paris in 1801. The pattern was determined by a seried of stiff cards with holes punched in them. During weaving a series of rods carry the different coloured threads into the loom, and the job of the punched card was to block some of these rods and let others pass through to complete the weave. The distribution of the holes on the cards determined the pattern on the silk, and the cards thus constituted a form of program (the spelling 'program', rather than 'programme', is used in computer circles). When Jacquard died, his portrait was woven on one of his looms using 24 000 cards. Babbage brought back one of these portraits for Queen Victoria.

Punched cards reached a zenith in 1886, when the American Hollerith used them in an electromechanical system for tabulating the 1890 American census. Holes were punched one at a time as the census forms were read, breaking down the data into simple yes/no answers. Fig. 5.6 is based on a drawing in Hollerith's patent application of 1889. The resultant punched cards were passed over a roller, inside of which were mercury containers, one for each row of holes. When a hole was present a probe was able to pass through and make electrical contact with the mercury, and this sent an impulse to the tabulator's counting device. The 1890 census was completed in six weeks, compared with the six years for the 1880 census. The population of the USA then was 62 622 250.

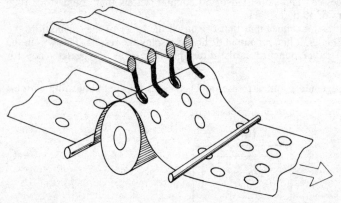

Fig. 5.6. Hollerith's electromechanical system for reading punched cards.

5.4   The Electronic Computer

The work of Babbage and Jacquard laid the foundations for modern computing, but it was not until the advent of electronics that their ideas were able to be realised. The major milestones are listed:

1939–44. Aiken of Harvard University worked with IBM engineers to produce an electromechanical calculator with punched card input. This was the Harvard Mark 1.

1946. ENIAC (Electronic Numerical Integrator and Calculator) was developed by Eckert and Mauchly of the University of Pennsylvannia. It was electronic, all-purpose, contained around 10 000 vacuum tubes and consumed 150 kW. It was a thousand times faster than the Harvard Mark 1.

1947. Von Neumann proposed that the program of instructions could be held in the store of the computer.

1948. Newman developed a stored program computer at Manchester University.

1949. A stored program computer EDSAC (Electronic Delay Storage Automatic Computer) was built at Cambridge University. It could add in 1500 microseconds.

Computer development has accelerated rapidly since then, and we now have the third generation of computers. The first generation, such as EDSAC, used vacuum tubes for the electronic components, and electrostatic tubes or mercury delay lines for storage. The second generation came along around 1957 when the first transistorised computers came on the market. This allowed a great reduction in size. They used magnetic drums and magnetic cores for storage. The third generation originated in the late 1960s, when discrete electronic components were replaced by integrated circuits.

The latest generation is vastly superior to the first. Compare ENIAC with the modern Fairchild F8. ENIAC consumed 56 000 times more power; at 30 000 kg, its mass was 60 000 times greater; it was 10 000 times less reliable; its volume was 300 000 times greater. And in spite of all this the price of the F8 is much less than that of the ENIAC.

5.5   What is a Computer?

A computer is an **information-processing** machine. It takes in information, processes it, and provides an output, based on the information and the processing.

There are two basic types of computer: **analogue computers** and **digital computers**. In the analogue computer physical quantities (usually voltages) are used to represent the variables in the problem under investigation. Use is made of the physical analogy of the equations describing the problem and its representation on the computer. For example, one can construct an electronic circuit in which the voltages behave in a similar manner to the displacements of a car suspension system. This is possible because the equations describing both systems can be written in a similar form. So the variations in voltage within the electronic analogue can give a measure of the variation of the displacements of the suspension. The voltage varies in direct proportion to the displacement. It also varies in a continuous manner, so that for any position of the suspension there will be an equivalent value of voltage.

We shall meet the analogue computer again in Chapter 6, so let us move on to the digital computer which is the main concern of this chapter. Unlike the analogue, the digital computer deals in discrete quantities. There are no smooth continuous operations in the digital computer, and we shall see more of this later.

The information in the output from the digital computer can be in **numeric form** (numbers) or **non-numeric form** (letters of the alphabet and punctuation). We should emphasise that doing sums is only one of the many tasks the computer has to face up to today. Updating last year's telephone directory, making out payslips, booking seats on an aircraft and translation are some typical tasks that use non-numeric data.

The basis of operation of a modern computer can be understood by comparing it with our own method of processing information. Referring back to Fig. 5.5, you will see that there are five basic operational elements: the input device, the output device, and the central processing unit (CPU) consisting of a memory, a store and an arithmetic logic unit (ALU). Imagine that I asked you to work out an examination problem in physics. The CPU consists of you, a pocket calculator, a pencil and a scribbling pad. The input device is the examination paper. The output device is your examination script. Within the CPU we can identify three basic elements. Obviously the pocket calculator is the ALU in this case. Whilst carrying out calculations, you would remember some intermediate results and you would note down some others on the scribbling pad. So the pencil and pad and your brain are the memory unit. And finally, the control unit is your goodself, for you have to organise the sequence of the calculations and the presentation of the output.

It would be appropriate to introduce some more jargon at this stage; **hardware** and **software**. Hardware comprises the physical bits and pieces that go to make up the computer. Software comprises the sets of instructions or program that control the operation of the hardware. So in our example, the pencil and paper and calculator would be the hardware, whilst the software would be 'all in the mind'.

5.6   Data Representation in the Computer

A computer is not simply a collection of individual electronic units. The intercommunication of the units is important, so it is essential that there is some standardised language for communicating information from one unit to another. The basis of this language is the computer's ability to recognise **two states**—current *on* and current *off*. In such a **binary** system all the information has to be represented by two states, or two symbols. Such a concept is not entirely new; it was used by Morse as the basis of his famous code. The two symbols in that case were the dot · and the dash – . But the computer, rather than using the dot and the dash, which are actually short and long pulses of electricity, uses the presence or absence of pulses, and calls them 1's and 0's respectively. So a sequence of 'pulse, no pulse, pulse, pulse' would be interpreted as 1011. Such a system can be implemented quite easily in a computer. Opening and closing of a switch can represent the two required states, and this can be done extremely quickly nowadays (about 1 ns) using transistors. The absence or presence of a hole in a punched card can be interpreted as an 0 and a 1 by the computer.

These 0's and 1's are referred to as **bits**, short for binary digits. Just as we express information in words, so do computers. A computer **word** is a group of bits, and the length can vary from computer to computer: some words are as short as 8 bits, others as long as 60. In many computer systems a set of 8 bits is called a **byte**, and can be used to designate numerical digits, numerals and many other useful symbols.

We can use a single bit to represent two symbols or characters. If its value is 0 we could arrange for the computer to interpret this as the letter A; if a 1, this could be interpreted as B. So the computer code is 0 = A, 1 = B, and you will recall that this means that the absence of a pulse means A, and the presence of a pulse means B. If we increase the word size to 2 bits, the code can handle four symbols, e.g. 00 = A, 01 = B, 10 = C, 11 = D. Extending this pattern you should be able to see that in general the number of symbols that can be represented is 2^N, where N is the number of bits in a word. For example, if $N = 5$, we can represent $2^5 = 32$ symbols, and this is enough to cover the whole alphabet, with some left over for punctuation, etc. The standard byte, with its 8 bits, allows for $2^8 = 256$ combinations.

5.7 Calculating in Binary

As well as being able to represent the letters of the alphabet, numbers and so on, the binary system allows us to do calculations with numeric data. We are used to calculating in the decimal system and we know, for example, that the number 2753 represents the sum of

2 thousands or 2×10^3
plus 7 hundreds or 7×10^2
plus 5 tens or 5×10
plus 3 units or 3×1

So $$2753 = (2 \times 10^3) + (7 \times 10^2) + (5 \times 10) + (3 \times 1)$$

The decimal system has ten symbols, 0 to 9, but the binary system has only two, 0 and 1. But binary numbers are constructed in the same way as decimal numbers. For example, the binary number 1011 represents the sum of

1 eights or 1×2^3
plus 0 fours or 0×2^2
plus 1 twos or 1×2
plus 1 unit or 1×1

So $$1011 = (1 \times 2^3) + (0 \times 2^2) + (1 \times 2) + (1 \times 1)$$

If you work out the right-hand side you will see that binary 1011 equals decimal 11. Table 5.1 gives the binary equivalents of the first 32 decimal numbers. You will note that it requires a 5 bit word to cover this range.

Table 5.1

Decimal	Binary	Decimal	Binary
0	00000	16	10000
1	00001	17	10001
2	00010	18	10010
3	00011	19	10011
4	00100	20	10100
5	00101	21	10101
6	00110	22	10110
7	00111	23	10111
8	01000	24	11000
9	01001	25	11001
10	01010	26	11010
11	01011	27	11011
12	01100	28	11100
13	01101	29	11101
14	01110	30	11110
15	01111	31	11111

It will be seen from the table that the price paid for representing information in the simple on-off binary system is that for a given number, a good many more binary digits are needed than decimal digits. This is summarised in the limerick:

He swears that the girls can't resist him,
Keeps a list of the ones who have kissed him.
The amount's not too hot
But it looks like a lot
Since its kept in the binary system.

Since the computer is based on the binary system then it is obvious that binary arithmetic has to be used for addition, subtraction, multiplication and division. Binary addition is very simple. There are only four possible outcomes for the addition of two bits. These are: $0 + 0 = 0$, $0 + 1 = 1$, $1 + 0 = 1$, and $1 + 1 = 10$. Note that the sum of 1 and 1 is 10 (equivalent to decimal 2), indicating a zero sum with a carry of 1. Let us look at a few examples of addition:

	Binary		Decimal
	1011		11
	101		5
Sum	10000	Sum	16
	110101		53
	100111		39
Sum	1011100		92

Binary subtraction is performed by reshaping the calculation into an addition using the method of **complements**. Let us see how this works for decimal numbers. In this case we use 'nines complements' which when added to a number raise it to 9 or 99 or 999 and so on. For example, the nines complement of 2 is 7, of 17 is 82, of 261 is 738 and so on. Using this method for subtraction requires the following rules to be followed:

1. Find the nines complement of the number to be subtracted.
2. Add this to the number from which you are taking away.
3. If the addition produces a carry, add 1 to the total;
   if the addition does not produce a carry, take the nines complement again and put a minus sign before the answer.

For example, conventionally

	37		12
	− 18		− 68
Answer	19	Answer	− 56

But using the complements

	37		12
	+ 81 (complement of 18)		+ 31 (complement of 68)
	(1)18		(0)43
	1 (add 1)		56 (complement of 43)
Answer	19	Answer	− 56 (add minus sign)

Now this is a very simple procedure to follow when the binary system is used, for since there are only two digits 0 and 1, we use 'ones complements' instead of the 'nines complements'. This means, in effect, that if we want to obtain the complement of a binary number, we merely change the 0s to 1s and vice versa. For example, 010101 is the complement of 101010. Let's see how it works by carrying out the binary equivalent of subtracting 33 from 59, i.e. 100001 from 111011:

	111011
	+ 011110 (complement of 100001)
	(1)011001
	1 (add 1)
Answer	11010 (decimal 26)

Multiplication and division are also carried out by converting them into addition processes, but not, I hasten to add, by continued addition like that carried out on earlier mechanical calculators where 3.3 × 1247 would be determined by adding 3.3 to itself 1246 times. This would be time-consuming even on the fastest of computers, so different and faster techniques have been developed.

5.8   The Computer's Basic Operations

We have spent some time on binary arithmetic, and now it would be interesting to learn how the bits and pieces within a computer actually carry out these basic operations. But before we can appreciate that, we must have an understanding of logic gates.

Logic Gates

The computer applies the principles of **logic** when carrying out numerical calculations. But you may wonder what Aristotle's formal logic has to do with calculation when it was originally designed to help us to reason out problems of logical truth. The English mathematician and logician, George Boole (1815–64), played a leading part in the development of a mathematical logic that laid down mathematical rules for combining statements that would yield logically valid conclusions. This technique, Boolean algebra, expressed logical relations by means of mathematical symbols. It was known as a **symbolic logic**.

It was not until 1937 that Claude Shannon realised that Boole's symbolic logic could be used for the design of electrical switching circuits. He showed that Boolean algebra could be used to determine the effects resulting from various combinations of switches. Consider, for example, Fig. 5.7(*a*) which shows two switches A and B in series with a battery. Each switch can have two states (they are binary devices), and we can designate the open state by an 0 and the closed stage by a 1. Shannon wanted to know how the flow of current around the circuit was related to the various possible combinations of switch positions, and the table in Fig. 5.7(*b*) helped him to resolve that problem. It shows the four possible switch combinations and the current flow relevant to each combination. For example, if either switch is open (A or B = 0), then no current will flow and the lamp will not be lit. This is designated by an 0 in the current column. Current will only flow and the lamp will light (X = 1) when both switches are closed (A = B = 1). In summary X will only be 1 when both A AND B are 1. The AND is in capital letters because it indicates the use of AND logic, which is one of the basic elements of Boolean algebra. Using the language of Boolean algebra, we can describe the behaviour of our switching current by the equation

$$X = A \ AND \ B$$

This very elementary example helps us to understand how Boole's work can help in circuit design. Shannon recognised the similarity between our conclusion that X = 1 when A AND B equal 1, and Boole's more general logic statement about AND logic. For example when Boole wrote *p* AND *q*, he meant *p* and *q* to be two statements each of which could be true or false, and that *p* AND *q* would be true if, and only if, both *p* and *q* were true. You will note that, in logical terms, each statement is of a binary nature—it can be true (1) or it can be false (0). And pursuing the analogy further, it is interesting to note that tables like those in Fig. 5.7(*b*) are known to this day as **truth tables**.

A device that implements AND logic is known as an **AND logic gate**, or

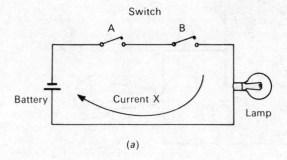

(a)

A	B	X
0	0	0
0	1	0
1	0	0
1	1	1

(b)

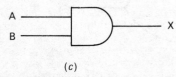

(c)

Fig. 5.7. AND logic: (a) using series switches; (b) truth table; (c) symbol for a two input AND gate.

in shortened form, as an AND gate. Its standard symbol is shown in Fig. 5.7(c). In practice today's computers do not use the slow solenoid-operated switches that Shannon used. The transistor (see Chapter 4) is the building block of the modern computer's logic, and **TTL** (transistor-transistor logic) is commonly used to build basic logic gates and circuits. There are many different types of logic gate but, in addition to the AND gate, it is sufficient for our present purposes to study only a further two, the OR gate and the inverter.

The OR gate is illustrated in Fig. 5.8(a). Common sense tells us that the lamp will light if either switch A or B is closed, and this is confirmed in the truth table of Fig. 5.8(b). The logic of the OR gate is described by the equation

$$X = A \text{ OR } B$$

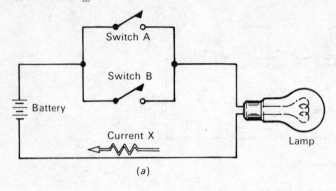

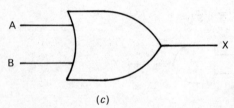

A	B	X
0	0	0
0	1	1
1	0	1
1	1	1

(b)

A ─────

B ─────

─────── X

(c)

Fig. 5.8. OR logic: (a) using switches in parallel; (b) truth table; (c) symbol for a two input OR gate.

Finally we look at the inverter, which is a contrary sort of gate, in that it gives an output X when there is no input A and vice versa. It is also known as a NOT gate (Fig. 5.9). A transistor can operate as a NOT gate where the output voltage drops to a low level when the control voltage is turned on, and vice versa. The logic of the NOT gate is given by the equation

$$X = NOT \ A$$

Addition using Logic Gates

We now know enough about logic gates to enable us to understand how the all-important addition operation can be carried out in the computer's arithmetic logic unit. When a pair of bits A and B are added, there are

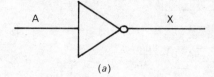

(a)

A	X
0	1
1	0

(b)

Fig. 5.9. NOT logic: (a) symbol for NOT gate; (b) truth table.

only four possible outcomes. These can be shown in truth tables as follows:

Inputs		Outputs	
A	B	Sum S	Sum C
0	0	0	0
0	1	1	0
1	0	1	0
1	1	0	1

You will note from this that the Carry C obeys AND logic, C = A AND B. The logic for the sum S is, however, not so obvious, and Fig. 5.10(a) shows how it can be built up from the basic gates we have just discussed. In Fig. 5.10(b) the particular case of A = B = 1 is considered, and the heavy lines indicate those that are carrying current. As expected, for this case, the Carry C will be on, and the Sum S will be off. You should check out the other three input combinations.

The device in the Fig. 5.10 is known as a **half adder**, and the reason for this becomes clear if we consider a binary addition. For example

$$001$$
$$+\ 011$$
$$\overline{}$$
Carry (0110)
Sum 0100

The first step in the addition, 1 + 1, can be carried out by a half adder with A = B = 1. But the second stage requires the addition of 0 and 1

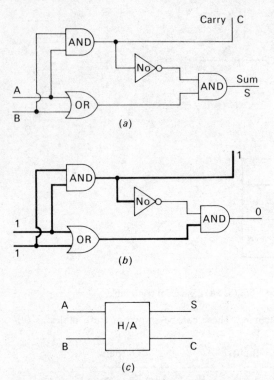

Fig. 5.10. (a) The half adder. *(b)* The situation when A = B = 1. *(c)* Symbol for the half adder.

(A = 0, B = 1) *and* a carry of 1 from the first half adder. So 3 bits have to be added, whilst the half adder can only cope with two inputs. It takes two half adders to carry this out, hence the name half adder.

Fig. 5.11 shows how five half adders can be used to add the two 3 bit numbers mentioned above: 001 and 011. The combination of two half adders and an OR gate constitutes a **full adder**, so the circuit actually consists of two full adders and a single half adder. You should check the circuit for different values of the input numbers. It is possible to extend the circuit to the left to allow binary integers of greater length to be added. Every additional digit would require the use of a further full adder.

5.9   Control

Now that we are reasonably familiar with the wonders of logic it would be an appropriate point at which to consider some other aspects of the computer's operation that require the use of logic. Let us move from one area of the central processing unit, the arithmetic logic unit, and examine the mysterious world of its neighbour, the control unit.

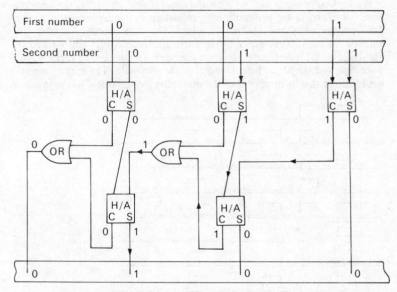

Fig. 5.11. Five half adders and two OR gates are needed to add two three-bit numbers.

When a word is moved around within the computer, its constituent bits are moved in **parallel** rather than in **series**. This means, for example, that all the bits of 1001 leave and arrive together, rather than going off one after the other and arriving in the same order. This requires as many parallel wires as there are bits in the word. Fig. 5.12 shows how a series of 5 bytes can be moved along eight parallel wires. A 1 bit is represented by the presence of a pulse and an 0 bit by the absence of a pulse. The destination is to the right of the figure, and whether it be a printer, a store or whatever, the bytes will arrive in the order 11001010, 01101011, 11010110, 10011011, 10110101 (reading upwards).

A byte can carry data or instructions or both. We saw earlier how numeric data can be added and now we will learn how an instruction, in the form of a binary number is obeyed. Let us assume that we have an imaginary computer that can only handle 4 bit words and is designed to recognise the word 1000 as the instruction for adding and 1001 as the instruction for subtracting. How exactly does the computer manage to recognise these particular arrangements of bits? This is where logic gates can help. Fig. 5.13 shows how it can be done. The information arrives along the four wires at the top of the diagram (these read upwards as on Fig. 5.12). As shown, the **code word** for addition, 1000, has just entered and the resulting signal level at each logic gate has been noted. The result is that the switch to the addition box is closed. You should check that when the word 1001 enters, the subtraction box will be brought into action. By the use of NOT gates and AND gates as shown it is possible to recognise any of the 16 possible combinations of the bits in our 4 bit word.

We now have some idea of how the control unit interprets the instructions contained in the program. It is important to distinguish between the data circuitry and the control circuitry. The data circuitry carries the numeric information relevant to the particular problem. If you wanted to find the area of a rectangle, the data circuitry would have to carry words describing the length and the breadth of the rectangle. The control circuit has to direct this information to the multiplier, to the store, to the printer,

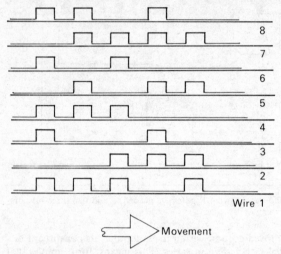

Fig. 5.12. Bits of information are moved in parallel. Here five bytes are carried to the right along eight wires.

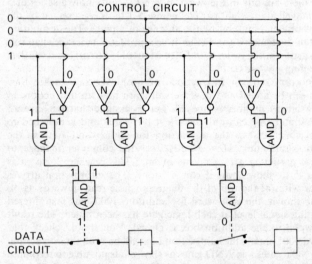

Fig. 5.13. Logic gates in a control circuit identify the code word 1000 for addition.

and so on, when required. As shown in Fig. 5.13, the control circuit does this by opening and closing switches in the data circuit.

Let us conclude this section on the control unit by finding out how information can be moved out of store when required. Every computer is controlled by a 'clock' that generates pulses at a constant rate; in most cases at several million per second. The time at which each event in a program takes place is related to the beat of this clock and is determined by the control circuitry. Indeed, the computer is a very complicated form of the sequence controllers we met towards the end of Chapter 4.

The **clock pulses** are used to move bits around the computer circuitry. Imagine that a 4 bit word is held in the computer store (we shall see how this is done later), and it is desired, for example, to move the word on to an adder. Fig 5.14 is a simplified diagram of how this is accomplished. Wires carry a clock pulse into each bit of the word, in this case 1010, and the system is so designed that the clock pulse can only pass through if the stored bit is 1. By this means the stored information is passed on.

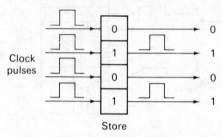

Clock pulses

Store

Fig. 5.14. Clock pulses moving information (1010) out of store.

But how did the clock pulses find their way to the correct part of the store in the first place? Clearly if the clock pulses had access to all parts of the store simultaneously then all the information would be passed on at one time and this would create havoc. There has to be a means for getting at the desired part of the store, or to use computer jargon, to find the correct **address** in store. Fig. 5.15 on p. 170 shows a switching circuit that can direct clock pulses to any one of 16 store addresses. Although the diagram shows single lines, there would in practice be four wires carrying pulses in parallel to each stored word. The control circuit recognises the address portion of the program instruction (1011 in the example) and causes the switches to be positioned up for an 0, and down for a 1. The clock pulses are then channelled to address 1011 and can move the contents on to another destination, as in Fig. 5.14.

5.10   The Store

Fig. 5.5 showed that the store is another major vital subsystem within the computer. The computer's basic processing cycle of input, process and output requires that data should be read into the main store from an input device (such as a keyboard) before it is processed. When the data, including

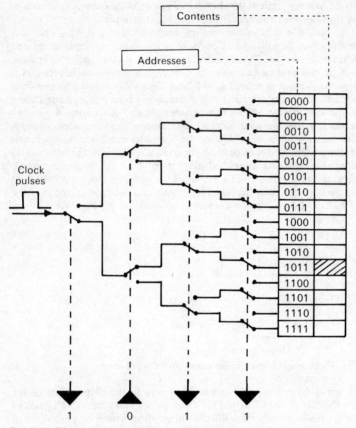

Fig. 5.15. Directing clock pulses to a particular address in store—here 1011.

the program, have been safely tucked away in this store, the computer can then start to process the information and produce an output.

The **main store** can be thought of as a neat array of little boxes, each with its own label or address, and each capable of holding one word. Clearly the more boxes there are the greater the capacity of the memory. The symbol K is commonly used to represent 1024 (2^{10}), so that a computer with an 8K memory would be capable of storing 8192 words. Many computer memories are designed to work with bytes (8 bits) and capacities are therefore often quoted in kilobytes and sometimes in megabytes (mega is a million). A typical microcomputer will contain between 4K and 64K bytes in main store. Minicomputer stores range from 64K to 1 megabyte, whilst the larger computers extend from 512K to 16 megabytes.

Most computers will require a **'backing' store** to extend the size of the main store, and we will discuss this in greater detail later. But for the present we focus on the main store which, in addition to having as large a capacity as possible, should be able to give up its information as quickly as possible. The time taken to retrieve data from memory is known as **access time**, and for today's computers this lies between about 50 ns and 1 μs.

But how does the computer store information? What do the little boxes consist of? Basically all the store has to do is to be able to distinguish between an 0 and a 1. You could, for example, imagine that each box in the store contained eight lamps, one for each bit in a byte. If you wished to store a byte you would then switch a lamp on for each 1 bit, and off for each 0 bit. The information would then be memorised and you could come back at a later time to 'read' the row of lamps. The lamp is one form of two-state device, on-off, that could be used for storage, but clearly it would be impractical in a computer. Magnetic cores, flip-flops and bubble memories are used in practice.

Magnetic Core

The magnetic core or ferrite core device was developed in the 1950s and was very popular until the 1970s when it was gradually replaced by semi-conductor stores (silicon chip memories). Like the lamp, the magnetic core is a two-state device. It is a small iron ring that can be magnetised, or polarised, in one of two directions (Fig. 5.16(a)). The polarity indicates whether the core is 'on' or 'off'. The magnetisation is a result of the magnetic field created by an electrical current passing through the centre of the

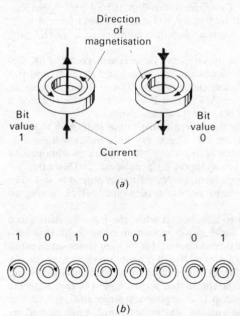

Fig. 5.16. Magnetic cores: (*a*) the direction of magnetisation, and the bit value, changes with the direction of current; (*b*) eight cores are needed to store a byte—here 10100101.

core. The direction of the magnetisation depends on the direction of this current. The current carries the data to be stored and when it stops flowing in the wire, the core retains its magnetism and hence 'remembers' which direction the last pulse of current was going in. So in order to store a 1 it is necessary to send a current pulse through the core in one direction, whilst for the storage of an 0, a current pulse has to pass in the other direction. A group of 8 cores is necessary to store 1 byte (Fig. 5.16(b)). In practice these cores are very small, down to 1 mm in diameter. They are strung on a matrix of wires, some used to carry data into the store, and some used for recognising the state of polarity of the cores—known as reading from store.

Semiconductor memory has reduced access time considerably and magnetic core is now only found on the oldest of machines. Access times have reduced from millionths of a second to thousandths of millionths of a second, and this has been accompanied by a surprising reduction in cost. In 1964 a megabyte of magnetic core memory could have costed around £1 million, whereas the same capacity in semiconductor form nowadays costs around £6000.

The Flip-Flop

At the heart of semiconductor memory systems lies the flip-flop, and we should therefore devote a little time to its study. It will put your understanding of logic to the test. One form of flip-flop, the SR or set-reset flip-flop is shown in Fig. 5.17. It uses two AND gates and four invertors. (In practice AND gate A_1 and invertor N_2 would combine to a NAND gate; similarly for A_2 and N_4.)

How does it work? Just like the lamp and the magnetic core, the SR flip-flop has two stable states. It has two inputs, S(set) and R(reset), and two outputs A and B. In one of these states A is on (1) and B is off (0); in the other the situation is reversed A = 0 and B = 1. The device is called a flip-flop because it flip-flops between these two states.

Fig. 5.17(a) shows that the flip-flop can exist in the state A = 0, B = 1 when both inputs S and R are off. The heavy lines indicate those that are carrying current. The other state, A = 1, B = 0, can be obtained by switching on the S input, i.e. by applying a pulse at S. This turns off N_1, which turns off A_1, which turns on N_2 giving an output A = 1. The feedback from this output turns on A_2, which turns off N_4 giving an output B = 0.

Now the important thing to note is that when the input S returns to 0 again, the outputs from the flip-flop will remain unchanged. All that happens is that the output N_1 is turned on, but this by itself is not sufficient to turn A_1 on again. So the state A = 1, B = 0 is maintained, and we can say that the flip-flop has 'remembered' that a logic 1 had been applied at input S. The circuit will remain in the stable state until a logic 1 is applied to the input R. Hence the SR flip-flop is a single-bit storage element and will remember, or store, a pulse entered via the set input, until cleared by entering a clear signal at the reset input. Thousands of such elements make up the computer's memory system.

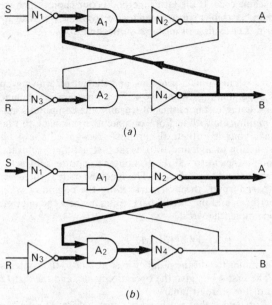

Fig. 5.17. The SR flip-flop: (*a*) B = 1 when S = R = 0, but (*b*) the outputs flip over when S = 1.

RAM and ROM

Two very common forms of memory are the RAM (**random access memory**) and the ROM (**read only memory**).

RAM is designed to allow access to any randomly chosen address to be achieved in a fixed time. So it does not matter where information is to be stored or where it is to be retrieved from—it will take the same time to get at it. You will see later that magnetic discs, used as back-up stores, have a similar property, whilst tapes do not.

Information can be read from ROM, but the computer cannot write information into it. Data is recorded in ROM when it is manufactured, so it is used to store programs that do not need to be altered, such as BASIC compilers used in microcomputers (see Section 5.11).

5.11 How to tell the Computer what to do

We use our native language to communicate with each other, and its large vocabulary allows us to convey a wide variety of information. But computers do not understand English: they only recognise 0's and 1's. So if we wish to instruct them we have two choices. Either we write our instructions as 0's and 1's, or else we provide some means for translating our instructions into that form. If we choose the former method, we will have to use what is known as **machine code**. If the latter method is our choice, then we will have to provide an 'assembler' or a 'compiler' to translate our instructions into binary form. Let us first of all look at machine code.

Machine Code

The list of instructions is known as a **program**, and the act of drawing up the list is known as programming. When we write a program in machine code, we ourselves are going to the bother of learning the computer's language in order to communicate with it. For example, referring back to Fig. 5.13 you will see that your hypothetical computer recognised the 4 bit word 1000 as the instruction to add, and 1001 as the instruction to subtract. But it is obviously not sufficient to tell the computer to add or subtract— we must also tell it what to add or subtract. The machine code is therefore presented in a two-part format, the first part telling the computer what operation it has to perform, and the second part telling it where the relevant data is stored. For example, the word

$$1000 \quad 000011010011$$

would instruct the computer to add the number in store 211 to the contents of the **accumulator**. The first 4 bits give the operation code, and the last 12 bits give the address of the store in binary.

The accumulator is a device that we have not met before. During the execution of a program there is a movement of information within the CPU and it is necessary at various stages to store information on a temporary basis. This is done in the accumulator.

The above instruction for addition is long, and you can well imagine that it would be a tiresome task to write a long program of such instructions. Fortunately, computer input devices are available to translate decimal into binary, so the above instruction can be rewritten in a much simpler fashion as

$$8 \quad 211$$

We can now propose a list of machine codes for our hypothetical computer. They each have a two-part format.

Table 5.2

01	000	* Clear the accumulator
02	000	*Stop
03	A	Read the next item of data and store it at address A (decimal)
04	A	Print the number stored at address A
05	A	Transfer a copy of the number in the accumulator to address A
06	B	* Enter the number B in the accumulator
07	A	Subtract the number in address A from the contents of the accumulator
08	A	Add the number in address A to the contents of the accumulator
09	A	Multiply the contents of the accumulator by the number in address A
10	A	Divide the contents of the accumulator by the number in address A

You will note that the instructions marked with an asterisk, whilst having a two-part format, are different from the others in that the second part does not refer to an address.

Let us apply this code to a simple problem. Given the length and breadth, in feet, of a rectangle, determine its area in square inches. We can express this as $Z = 144\ XY$. where Z is the area, X the length and Y the breadth. The program is as follows:

Table 5.3

Instruction number	Instruction		Mnemonic form	
1	01	000	CLA	
2	06	144	INA	144
3	03	201	INP	201
4	03	202	INP	202
5	09	201	MLT	201
6	09	202	MLT	202
7	05	203	TR	203
8	04	203	PT	203
9	02	000	END	

We will ignore the right-hand column for the moment. Let us go through this program, instruction by instruction, for the case X = 2 feet, Y = 3 feet. These data are the inputs to the program and must be supplied by a keyboard, a tape or punched cards. Fig. 5.18 assumes that cards are used for input, and examines the state of the store and of the accumulator at the completion of each instruction. The program uses the three addresses 201, 202 and 203, and we assume that at the start of the program, data, say 5, 112 and 62 are left over in these addresses from some earlier calculation. The first instruction clears the accumulator, setting it to zero. The second instruction enters the number 144 into the accumulator. The third instruction reads 2(X) into address 201—and so on until the area 864 is printed out.

Assembly Code

The construction of a machine code program can be a daunting task, even when written in decimal rather than binary. Our simple system uses only ten codes, but there can be many more in a real system, and it is a tiresome job to have to refer continuously to a long list to make sure that the correct numerical code for a given operation is being used. It would be useful to have some form of memory aid that would help us to write the program without reference to a list of numerical codes. **Assembly code**, or assembly language, is such a memory aid. It is sometimes referred to as a **mnemonic code**, a mnemonic being a memory jogger. For example, if the code CLA, for 'Clear the Accumulator' were used instead of 01, we would have a better chance of remembering it. We can suggest mnemonics for all ten of our computer's machine codes as shown in Table 5.4.

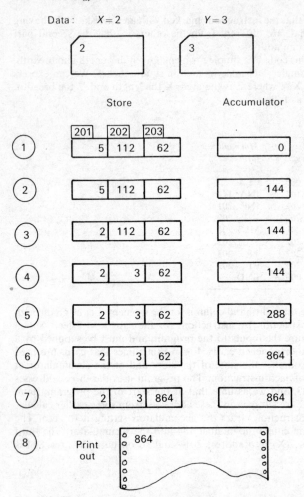

Fig. 5.18. The states of store and accumulator as our simple computer follows
through a machine code program.

If you refer back to Table 5.3, you will see the assembly code version of
the program in the right-hand column. Now, although the use of assembly
code makes life easier for us, it makes it a lot more difficult for a computer
that only understands 0's and 1's. So we have to translate the assembly
code back to machine code before the computer can respond. This is done
by an **assembler**, a program specially written for the job. For example,
the assembler would translate the assembly code MLT 202 into
1001 000011001010.

We have seen that machine code is directly 'understood' by the computer's circuitry, so since there are many different makes of computer with

Table 5.4

Machine code	Assembly code	Meaning
01	CLA	Clear accumulator
02	END	Stop
03	INP A	Input data to A
04	PT A	Print contents of A
05	TR A	Transfer to A
06	INA B	Input B to accumulator
07	SUB A	Subtract contents of A from accumulator
08	ADD A	Add contents of A to accumulator
09	MLT A	Multiply accumulator by contents of A
10	DIV A	Divide accumulator by contents of A

many different types of circuitry, it is easy to conclude that there must be many different machine codes, each specific to a particular make of computer. Because of this, **machine languages** and **assembler languages** are known as computer-orientated languages—a particular computer requires a particular language. They are also known as **low-level languages**.

High-Level Languages

Low-level languages are efficient in terms of use of storage, and they can run through a program very quickly. Nevertheless, there is a strong demand nowadays for high-level languages—languages that make programming easy, and that can be used on any computer. Such languages are called user-orientated, because they are more concerned with the user than the particular computer that is being used.

Many high-level languages have developed over the years. Of the 200 available we shall have to restrict ourselves to a brief look at four of them:

FORTRAN (FORmula TRANslation). FORTRAN was the first high-level language and was developed by IBM (International Business Machines) in 1954. It was designed for use by scientists, engineers and mathematicians, allowing them to write programs in a form that closely resembles the equations of their own discipline. A sample instruction could be

$$B = 1 + C*SIN(D/E)$$

You would not be surprised to learn that this represents the equation

$$B = 1 + Csin(D/E)$$

COBOL (COmmon Business Orientated Language). The many everyday commercial problems of the US Department of Defense led to the development of COBOL around 1958. It was soon found that FORTRAN was not suited to business problems such as processing the weekly payroll or assessing the efficiency of material flow through a factory—problems that require access to large files of information, including people's names and addresses, salaries, etc. COBOL was intended to simplify the programmer's task in such cases. It looks like English and is meant to be intelligible to those not skilled in programming. A typical instruction could be

MULTIPLY LOAN-AMOUNT-INPUT BY
INTEREST-RATE-CONSTANT
GIVING INTEREST-AMOUNT-WORK

PL/1 (Programming Language 1). We have seen that FORTRAN is science-orientated and COBOL business-orientated. Scientific problems usually require fast computational speeds, but there is not much need for large input/output operations. Business users, on the other hand, usually need access to a lot of data and this requires fast input/output operations. PL/1, developed in the mid-1960s, combined the best features of the earlier languages in the hope that it would be suitable for any kind of problem.

BASIC (Beginner's All-purpose Symbolic Instruction Code). BASIC is very easy to learn, and is commonly used for both scientific and commercial problem-solving. It was developed by Professors J. Kemeny and T. Kurtz in the mid-1960s at Dartmouth College in the USA. Because of its simplicity it nowadays finds widespread use. For example, the following three BASIC instructions are equivalent to the nine machine code instructions listed in Table 5.3:

Table 5.5

```
10   INPUT X, Y
20   PRINT 144 * X * Y
30   END
```

In addition to being much more compact and easy to write than machine code, this example shows that programming in a high-level language allows the programmer to concentrate on the actual problem; he is not diverted by problems of translation. It is not unlike the businessman who has to prepare a convincing marketing document for a potential German customer. He could attempt to write it directly in German—that is, solve the problem in a foreign language. Alternatively, he has a much better chance of doing a good job if he concentrates on the problem in his own language, produces a convincing 'sales pitch' in his own language, and then has it translated into German. The first approach is equivalent in computer jargon to using machine code. The second is equivalent to using a high-level language, such as BASIC.

Of course, once the program has been written in a high-level language, it has to be translated into the language that the computer understands—machine code. This is equivalent to translating the document in our example from English into German. The computer uses a special program called a **compiler** to do this. For example, a BASIC compiler would be used to translate the BASIC program of Table 5.5 to the machine code program of Table 5.3. Each high-level language requires a compiler and most makes of computer have their own sets of compilers. You should be able to see a similarity now between an assembler and a compiler. An assembler translates assembly code and a compiler translates a high-level language.

5.12   Programming Principles

In order to develop a greater awareness of the techniques and problems of programming we shall now devote some time to a closer examination of BASIC. Like all high-level languages, BASIC has a vocabulary—a list of key words and symbols—and a syntax—a set of rules or grammar for writing the words and symbols in correct sequence. Each statement (or instruction) in BASIC has to begin with a line number. In order to illustrate this let us return to the problem of finding the area Z of a rectangle in square inches, given the lengths of its sides X and Y in feet. Table 5.3 and Fig. 5.18 showed how machine code could be used to solve this problem. For the particular case X = 2, Y = 3, a BASIC program for this would be

Table 5.6

```
10   PRINT 144*2*3
20   END
```

The numbers 10 and 20 are the line numbers, and it is usual to increase these in tens, so that there is room to insert additional intermediate line numbers later if required. Two different statements of BASIC appear in this simple example, the PRINT statement and the END statement. The arithmetic operation of multiplication is signified by the symbol *. The intention of the program is clear: at line 10 the computer is instructed to evaluate $144 \times 2 \times 3$, and to print the answer; at line 20 the program ends. In practice using a microcomputer, line 10 would be entered at the keyboard, the carriage return would then be operated, line 20 would be entered and the carriage return once again operated. Then, in order to make the computer act, it is only necessary to enter RUN, and operate the carriage return. The answer, 864, would then be displayed on a screen or printed on paper.

Since we are dealing with a language, it is essential that the correct words and grammar are used. For example, the computer would not run if any of the following versions of line 10 were used

```
       10   PRANT 144*2*3
 or 10   PRINT 144 MULT 2 MULT 3
 or 10   PRINT, 144*2*3
 or 1B   PRINT 144*2*3
```

BASIC Arithmetic

The rules of arithmetic as they apply to BASIC are very similar to the normal rules. Addition, subtraction and division use the usual symbols. Multiplication, as shown above, uses an asterisk, and exponentiation or raising to a power, uses a vertical arrow. For example 2↑3 means 2^3. When carrying out an involved arithmetical operation it is essential that the correct order of the individual operations is used. For example, does $2 + 6*3$ mean $(2 + 6) * 3 = 24$ or $2 + (6 * 3) = 20$?

An order of priority has been established for BASIC, and it is essential that it be adhered to. It is, starting with the top priority

    ( )   quantities in brackets
    ↑   exponentiation
    * /   multiplication and division
    + −   addition and subtraction

So, in the above example, $2 + 6 * 3$ would be taken to mean $2 + (6*3)$, not $(2 + 6) * 3$. Multiplication has priority over addition. The order of solution of a more complicated case would be

$$2 + 6 * 3 \uparrow 2 - 5 + (12/2)$$
$$2 + 6 * 3 \uparrow 2 - 5 + 6$$
$$2 + 6 * 9 - 5 + 6$$
$$2 + 54 - 5 + 6$$
$$57$$

5.13   Program Development

The simple program of Table 5.6 calculated the area of a rectangle with sides 2 feet and 3 feet long. It only works for that case. If we wanted to find the area of a 5 by 6 rectangle we would have to rewrite line 10 as

    10   PRINT 144 * 5 * 6

and for every other change of side, we would have to rewrite this line. This can be avoided by adding a new line to the program as follows:

Table 5.7

    10   INPUT X, Y
    20   PRINT 144 * X * Y
    30   END

When this program is input to the computer, and is RUN, a question mark will appear on the screen. The computer is asking for the values of X and Y. When these are entered, with a comma between, and the carriage is returned, the answer for those particular values of X and Y will be displayed. So the addition of one instruction makes the program much more versatile.

Now we know what the program in Table 5.7 is intended to do, but perhaps in a few month's time we will have forgotten, or maybe we would wish someone else to use the program then. So there is a need for some means of making the program more easily understood. This can be done by the use of REM (for REMark) statements that allow explanatory remarks to be put into the written program. The PRINT statement can also be expanded to allow for greater clarification. Our program could be modified as follows:

Table 5.8

```
10   REM      RECTANGLE AREA SQ. IN.
20   PRINT    "GIVES 144 XY, WITH X,Y IN FEET."
30   PRINT    "TYPE IN X,Y"
40   INPUT    X,Y
50   PRINT    'AREA IN SQ. IN. IS"
60   PRINT    144*X*Y
70   END
```

When this has been typed in at the keyboard and RUN, the screen will display the following:

```
            GIVES 144 XY, WITH X,Y IN FEET
            TYPE IN X,Y
            ?
```

When we respond by typing X and Y, say 2 and 3, and operate the carriage return, the computer will proceed with lines 50, 60 and 70, and the display will finally read

```
            GIVES 144 XY, WITH X, Y IN FEET
            TYPE IN X,Y
            ? 2,3
            AREA IN SQ. IN. IS
            864
```

Note that the REM statement is not included in the output. However, if this program were stored on tape (see later) and at some future date it was read into the computer, then on typing the word LIST and operating the carriage ·return, all seven instructions would be listed, including the explanatory REM statement.

A further development of the program could result in the following:

Table 5.9

```
10   REM   RECTANGLE AREA SQ. IN.
20   REM   SIDE OF EQUIVALENT SQ.
30   PRINT "GIVES 144 XY, WITH X,Y, IN FEET"
40   PRINT "AND SIDE OF EQUAL SQ. IN FT"
50   PRINT "TYPE IN X,Y"
60   INPUT X,Y
70   IF X > 25 THEN 160
80   IF Y > 25 THEN 180
90   Z = 144*X*Y
100  S = SQR(Z)
110  PRINT "AREA IN SQ. IN. IS"
120  PRINT Z
130  PRINT "SIDE OF EQUAL SQ. IS"
140  PRINT S
150  END
160  PRINT "X IS TOO BIG"
170  END
180  PRINT "Y IS TOO BIG"
190  END
```

This program differs from that of Table 5.8 in several ways. First, in addition to finding the area of the rectangle in square inches, it also works out the side of the square that has the same area. This is why line 100. $S = SQR(Z)$ is included. BASIC has a built-in library of commonly used functions like this, sincluding SIN, COS, TAN and LOG.

Another major change in the program is the introduction of an element of **decision-making**. You will recall that, in the introduction to this chapter, we noted that decision-making was one of the basic functions of the brain that has been extended by the computer. Our program is concerned with the calculation of the area of a rectangle, and it could be used, for example, by an architect to determine the area of a room. Now let us assume that in building this room the architect requires a continous wooden beam along each wall, and let us further assume that the maximum available length of beam is 25 feet. Therefore it is clear that he cannot consider rooms for which X or Y is greater than 25 feet. This is the reason for the insertion of lines 70 and 80. Line 70, for example, states that if the side X exceeds 25 feet in length, then the program should jump to instruction 160 which tells the computer to print "X IS TOO BIG". Lines 80 and 180 deal with Y in a similar manner. This ability to compare or test quantities in order to decide the course of a calculation is a fundamental and important facility of BASIC. It is at the heart of the decision-making process.

Flow Charts

Our example program has been developed in a somewhat haphazard way.

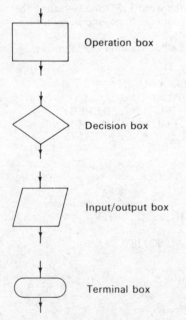

Operation box

Decision box

Input/output box

Terminal box

Fig. 5.19. Basic blocks for use in flow charts.

Rather than adding bits here and there as we went along it would have been more sensible, and more profitable, to have considered the objectives of the program in a more logical and systematic way right from the outset. This is what flow charts are for. A flow chart is a blueprint for the program. It breaks down the solution of a problem into a sequence of clear logical steps.

In drawing flow charts we have to follow certain rules and conventions. Some of the more commonly occurring symbols are shown in Fig. 5.19. The operation or processing box is used every time information has to be processed, e.g. multiplied by, added to, etc. The decision box allows us to include questions to which the answer can be 'yes' or 'no'. The input/ output box is used to indicate those points in the program where information has to be read in or read out. And finally the terminal box, i.e. the start/stop box, is used to make it clear when our set of instructions begins and ends. Fig. 5.20 uses these basic symbols to construct a flow chart for the rectangle problem.

In concluding this section, it is appropriate to draw attention to the mysterious **algorithm**. Believe it or not, in drawing up the flow chart of Fig. 20 we have constructed an algorithm for our problem. An algorithm is simply the reduction of the problem into a series of steps in some particular sequence. It is a recipe. More formally, it is a finite unambiguous list of instructions specifying a sequence of basic operations, such that if the instructions are obeyed and the operation carried out in the exact order stated, then a solution will always be obtained to any particular problem belonging to a given class of problems.

5.14 Peripherals

We shall conclude this chapter by looking briefly at some of the computer's important peripheral equipment; in particular that associated with the store, and the input and output.

Input Devices

How do we communicate with the computer? We have seen earlier how the computer can be told what to do my means of programs in high or low languages. But how do we actually get these instructions into the computer? It would be nice if we could talk directly to the computer, using normal speech, and indeed a lot of research is going on in this field of speech recognition. Some progress has been achieved and devices are available that can recognise particular words. But the dream of direct speech control is still a long way from reality, and in the meantime we have to make do with less exciting input devices like punched cards and tapes.

The **punched card** is a development of Hollerith's invention of 1889 (see earlier). It contains data in the form of small rectangular holes punched in a card 187 mm × 83 mm. Cards (see Fig. 5.21(b)) are normally divided into 80 columns and 12 rows, each column representing an alphanumeric (a letter or number) or a special character.

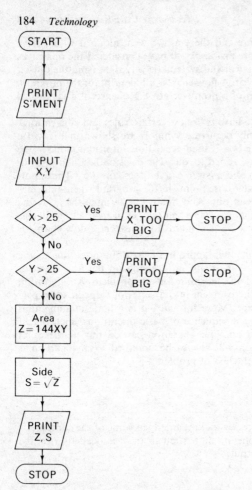

Fig. 5.20. A flow chart for the program in Table 5.9.

All letters require two holes, each number requires one hole, and special characters use two or three holes. The 80 characters represented by the 80 columns can provide a lot of information. For example, for identification of people, columns 1 to 12 could be used for the name, columns 13–40 for the address, and so on.

The card reader is the interface between the punched card and the computer. It normally uses a light source and a set of photoelectric cells; when a hole is present the light gets through and activates the relevant photo cell. Cards can be read at speeds varying from 300 to 2000 cards per minute, giving a maximum character transfer rate of about 160 000 per minute. Although this may seem impressive, it has to be remembered that the CPU can process information at about a thousand times faster. So card reading is a relatively slow process that does not take advantage of the capabilities of the fast modern computer.

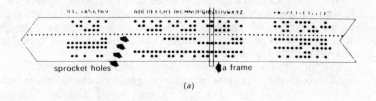

(a)

(b)

Fig. 5.21. Inputs to the computer: (a) punched tape; (b) punched card.

Punched paper tape is another well established form of input to the computer (Fig. 5.21(a)). Information is stored on the tape in the form of a line of holes across the width of the tape for each character. The tape shown in the figure is called 8-channel tape, because there is room for eight holes across the tape. There is in addition a smaller hole, the sprocket hole, used for driving the tape. For example, taking the presence of a hole to indicate binary 1, and reading from the bottom of the tape to the top, the code 0 101 0011 represents the letter S. The first bit of this sequence is known as the **parity bit**, and the remainder is written in ASC11 (American Standard Code for Information Exchange). The parity bit is used for the detection of errors that may occur during the punching or the reading of the tape. In a so-called 'even parity' system, the parity bit is used to ensure that the total number of holes representing any character is even. If you check Fig. 5.21(a), you will see that every character has an even number of holes. If, for example, the sixth bit was incorrectly punched for the letter S so that the holes read 0 101 0111, then the total number of holes would be odd, not even, and this would be interpreted as an error.

A typical paper tape is about 25 mm across and comes in reels about 300 m long. About 400 rows of characters can be punched on every metre of tape. Like the punched card reader, the paper tape reader uses an optoelectronic reading head. The tape moves past this at speeds up to 2.5 m/s, and at 400 characters to the metre, this gives a reading speed of about 1000 characters per second, which is slower than a card reader.

Tape is cheaper than cards and is easier to store. It does, however, have the reputation of being more difficult to verify and correct. Unlike cards it has the advantage of retaining its order after being accidentally dropped on the floor, but against this it is difficult to correct or to add to or delete from without cutting and splicing.

In addition to tapes and cards, many computers are now expected to be able to recognise marks and characters. **Magnetic ink character recognition** (MICR) uses highly stylised character shapes printed in an ink containing magnetisable particles. These shapes are read by a detector that produces an internally coded representation in the computer. The odd-looking numbers printed along the bottoms of cheques are examples of MIRC.

Optical character recognition (OCR) uses light instead of magnetic fields to sense the shape of the characters. The character is scanned by a spot of light and the variations in brightness generate a pattern of electrical pulses. This pattern is compared against a set of patterns, one for each character, stored in the computer. If it matches one of them, then that character is taken to be the correct one. If you have ever received an invoice or a subscription renewal form from a magazine or book publisher, it is highly probable that your name, address and code number were presented in OCR form. And the next time you buy something at the supermarket, have a close look at the odd array of parallel lines on the side of the packet. This is a bar-code, as on the back cover of this book; another example of OCR.

Finally, in our discussion of input devices, we should not overlook the **keyboard.** All of the inputs we have discussed so far require the data to be prepared by some separate device, e.g. a card punch or a bar-code printer. A keyboard allows us to communicate directly with the computer. With mini and microcomputers the keyboard is usually attached to the computer, but for the large main-frame computers it is not unusual to have remote operation by keyboard printer terminals, sometimes known as teletype machines.

Output Devices

The computer usually presents its information in the form of the printed word or as a picture or graph. The teletype machines referred to above can print the output from the computer as well as the input to it, but they are slow.

The **line printer** is one of the many devices that were developed to allow a much faster production of the printed word. The line printer can use either a drum or a chain to carry the embossed characters. Fig. 5.22 will help to explain how the **drum printer** works. The main part is a rotating stainless-steel drum in which the characters are etched in relief. A printed line normally consists of 120–136 characters, and there is a complete set of characters (usually 64) around the circumference of the drum at each of these print positions. The drum rotates at high speed but the paper is at rest during printing. As the drum rotates a hammer strikes the desired character at each print position. For example, when the row of A's is in the print position, magnetically operated hammers are actuated at all those positions where the letter A is required. The drum then rotates to bring the

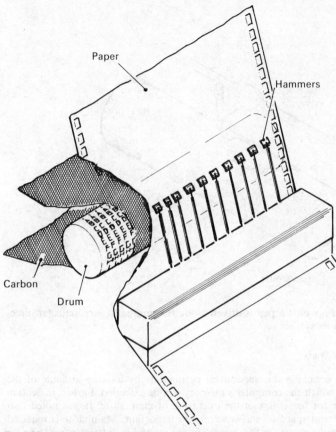

Fig. 5.22. A drum printer.

row of B's into position, and all the required B's are then printed. This continues until the drum has completed one full revolution, at which point a full line will have been completed. The paper is then moved on to the next line. Printing speeds around 600 lines per minute are typical.

The line printer gives a **hard copy**, i.e. a permanent record of the output. But there are many occasions when a hard copy is not really necessary, and in such cases a **visual display unit** (VDU) is sufficient. A VDU is a screen, driven by a cathode-ray tube, that is capable of presenting alphanumeric characters. Many microcomputers have a 'built-in' VDU, but most home computers use the ordinary TV screen for this purpose.

When a graphical output is required the VDU can also be used, but for detailed hard copy it is usually necessary to use a **graph plotter.** There are two basic types: the **drum plotter** and the **flat-bed plotter.** In the former, the drawing is constructed by rotating the drum backwards and forwards, whilst simultaneously driving an inked pen horizontally, from side to side (Fig. 5.23). In the flat-bed plotter, the paper is laid flat and is held

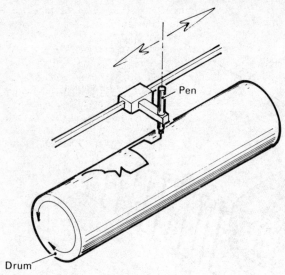

Drum

Fig. 5.23. A drum plotter.

stationary, whilst a pen is driven along two mutually perpendicular directions on its surface.

Backing Store

We will conclude this section on peripherals by looking at some of the ways in which the computer's memory can be extended. Earlier, in Section 5.10, we got some idea of the operation of main store. It was noted that capacity and speed of access were very important. Main store is required as a working space for the current program, and is only intended to retain information on a temporary basis. But many of today's problems require enormous amounts of data—much more than main store can handle—and it is necessary to provide some form of backing store that can retain these data.

In Section 5.5 we compared ourselves with computers and looked at the example of working out an examination problem. We identified the pocket calculator as the ALU, ourselves as the control unit, and our brain and the pencil and paper as the main store. Continuing with this analogy, we could consider backing store to be a reference book, a table of numbers or a filing cabinet of information that was relevant to the problem.

Magnetic tape and **magnetic discs** are common forms of backing store. The principle of operation of these devices is similar to that of the well-known tape recorder, but pulses representing patterns of bits are recorded instead of complex musical waveforms. Information is retained in the form of magnetised and non-magnetised spots (1's and 0's). Magnetic tape consists of a continuous strip of polyester plastic coated with magnetic oxide. Tapes are typically about 13 mm wide and about 700 m long. Recording densities range from 8 to 32 bits/mm, so a typical reel of tape can carry about 10 megabytes of information. Data can be transferred at rates between 20 000 and 300 000 characters per second.

A big disadvantage of magnetic tape is the fact that data are stored in a sequential manner, so if you want to get information from the end of the tape you have to wait until the reader has spun through most of the tape. The same problem exists with the common domestic tape recorder. A faster rate of access can be achieved by using direct access rather than sequential access. Magnetic discs provide this facility. In shape, a magnetic disc resembles an LP record, although there are no grooves in the surface. Discs are stacked in packs of six or more (Fig. 5.24) and they are mounted about

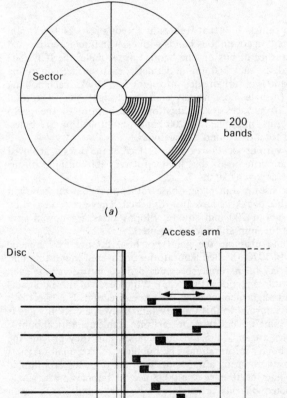

Fig. 5.24. Disc memory: (*a*) showing sectors and bands on an individual disc; (*b*) showing a stack of six discs with access arms to top and bottom faces.

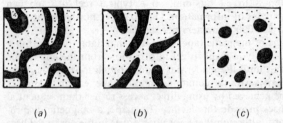

(a)   *(b)*   *(c)*

Fig. 5.25. Magnetic bubble memory: (*a*) to (*c*), increasing the magnetic field normal to the layer causes domains to contract.

12 mm apart on a central hub that rotates at speeds up to 2400 rev/min. Information is stored on continuous bands with each character represented by a patterned sequence of bits on one band. For example, the ICL 2802 disc store uses six discs, each 360 mm in diameter, each surface containing 200 bands. Each band is in turn divided into eight data blocks, each holding 512 six-bit bytes. (A byte more usually consists of eight bits.) The total capacity is then: 10 surfaces (excluding top and bottom of the pack) multiplied by 8 sectors, multiplied by 200 bands, multiplied by 512 bytes. This gives a grand total of around 8 megabytes.

Direct access to various locations on the surface of the discs is achieved by fast moving read/write heads that float on a very thin cushion of air. Access time can be as low as 20 ms.

The **'floppy' disc** uses a thin plastic base rather than the thicker rigid metal base of the disc packs we have just discussed. They are protected by a paper envelope about 200 mm square. Floppy discs are now a very popular peripheral for mini and microcomputers.

Finally, we should examine the great potential offered by magnetic bubble memories. In 1970, the Bell laboratories discovered that a thin film (10 μm) of garnet crystal naturally becomes divided into domains, each magnetised in opposite directions (Fig. 5.25). Within each of these domains all of the molecules align themselves in the same direction. If a bias magnetic field is applied normal to this thin crystalline layer, a critical value is reached when the domains suddenly contract into stable cylindrical bubble shapes. The bubbles are only 1–5 μm in diameter, and they provide the means for storing binary information. The bubbles are like small magnets so they can be moved about by a magnetic field; the presence of a bubble can be read as a 1 and the absence as an 0. Since the bubbles are so small, memories constructed on this principle can be very small for a given capacity. About 1500 bubbles can be accommodated in a square millimetre.

5.15   Exercises

Answers to numerical questions are given in Appendix 1.

1. Describe (about 50 words each) the contribution made to the development of computing by Napier, Pascal, Babbage, Hollerith and Boole.

2. Distinguished between analogue and digital computers. Draw up a list of pros and cons and give examples of applications.

3. (*a*) Define the terms bit, byte and word. (*b*) Express decimal 71 in binary. (*c*) Add 101101 and 110110; confirm your result in decimal. (*d*) Using 1's complements subtract the binary equivalent of 89 from the binary equivalent of 106. (*e*) Subtract binary 100110 from binary 010111 using 1's complements.

4. (*a*) A NAND gate is equivalent to an inverter in series with an AND gate. Draw its truth table. (*b*) Fig. 5.26(*a*) shows an arrangement of NAND gates called an EXCLUSIVE OR gate. Draw a truth table and, by comparison with Fig. 5.8, determine why this gate is so called.

5. Draw the circuit diagram for a half adder. Indicate which lines are 'live' for each of the four possible combinations of the two inputs.

6. Study the assembly of OR gates in Fig. 5.26(*b*). Can you see what it is designed to do?

7. Fig. 5.13 shows how the data circuit can be made to respond to the 4 bit words

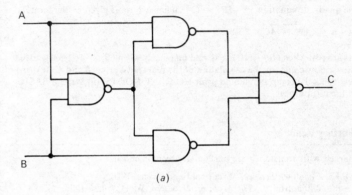

(*a*)

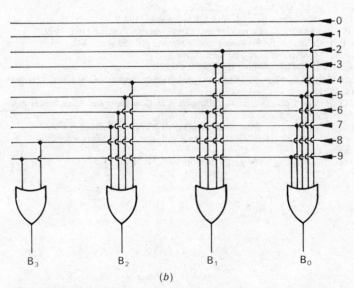

(*b*)

Fig. 5.26. (*a*) EXCLUSIVE OR gate (exercise 4(*b*)). (*b*) What function does this assembly of OR gates perform (exercise 6)?

1000 (add) and 1001 (substract) from the control circuit. Design an arrangement of gates that will respond to the instruction to multiply (see Table 5.2 for instruction number).

8. (*a*) Distinguish between main store and backing store. (*b*) Computer memories may be classified according to their capacity. Explain this classification, giving examples of the hardware and the associated access times.

9. Using Table 5.2, design a machine code program that will calculate the mean *z* of two numbers *x* and *y* and print the result. The two numbers are to be read from cards and the program should restrict itself to the addresses 109 to 112 inclusive.

10. (*a*) Distinguish between machine code and assembly code. (*b*) What is an assembler? (*c*) Write your program of Exercise 9 as an assembly code.

11. (*a*) Distinguish between high-level languages and low-level languages. (*b*) Give five examples of high-level languages. (*c*) Write a program in BASIC to calculate the mean *z* of two numbers *x* and *y*.

12. The quadratic equation $x^2 + Bx + C = 0$ has two roots given by the formula

$$x = 0.5(-B \pm \sqrt{(B^2 - 4C)})$$

The roots are real when $(B^2 - 4C) \geqslant 0$, and imaginary when $(B^2 - 4C)$ is negative.

(*a*) Draw a flow chart for the calculation of the real roots, given *B* and *C*. If there are no real roots the computer should print NO REAL ROOTS. (*b*) Write a BASIC program for this purpose.

5.16 Further Reading

Those marked with an asterisk are particularly recommended.

*Atkin, J. K., *Computer Science*, Macdonald and Evans, 1980.

Dale, R., and Williamson, I., *The Myth of the Micro*, W. H. Allen, 1980.

Day, C., and Alcock, D., *Illustrating Computers*, Pan Books, 1982.

Evans, C., *The Mighty Micro*, Victor Gollancz, 1979.

Fred learns about computers, Macdonald and Evans, 1981.

Hunt, R., and Shelley, J., *Computers and Commonsense*, Prentice Hall, 1979.

Maynard, J. *Computer Programming Made Simple*, Heinemann, 1983.

*Monro, D. M., *Basic BASIC*, Edward Arnold, 1978.

More Limericks, Bell Publishing Co. (NY), 1977.

Olsen, G. H., and Burdess, I., *Computers and Microprocessors Made Simple*, Heinemann, 1984.

Shelly, G. B., and Cashman, T. J., *Introduction to Computers and Data Processing*, Anaheim Publishing Co., 1980.

Strandh, S., *Machines, an Illustrated History*, AB Nordbok, Sweden, 1979.

The Computer: Yours Obediently, Scottish Computers in Schools Project, W. R. Chambers Ltd, 1975.

Trask, M., *The Story of Cybernetics*, Studio Vista, 1971.

6
Models and Optimisation: Tools for the Technologist

6.1 Introduction

Models are representations of reality. They are bodies of information about an element or a system collected for the purpose of studying that element or system. There is still no universally accepted classification of models, and you can read about descriptive models, predictive models, non-linear models, stochastic models, closed models and so on. We shall restrict ourselves to a close look at three main areas—**iconic** models, **symbolic** models and **analogue** models—and within these areas we shall distinguish between **static** and **dynamic** models.

6.2 Iconic Models

Iconic models look like what they represent; they are images. Two-dimensional iconic models would include photographs, drawings and maps, and there are also many three-dimensional iconic models, better known as **scale** models (Fig. 6.1). These can be scaled down, such as models of the planetary system, or scaled up such as Bohr's model of the atom, or Crick's model of the DNA molecule. Other examples are a scale model of a city which may be used during the planning of a new road layout or a scale model of an estuary which could be helpful in solving silting problems. The advantages of iconic models are obvious; clearly it is easier to handle a map of the country than the country itself, and it is much more convenient for traffic engineers to use a scale model of the city rather than the city itself.

It is necessary at this stage to introduce readers to the concept of **simulation**, and those of you not familiar with the technologist's interpretation of the word will need some explanation. The dictionary definition of simulation includes 'a profession meant to deceive—a counterfeit—a fraud', although when I asked a colleague for his definition, he replied: 'Margarine'. But let us agree to define simulation as experimentation with models. Thus modelling involves the representation of a system or component, and simulation involves experimentation with the model in order to find out how the real system or component will behave.

For example, simulation of an estuary would require a flow of water to be set up in the model, and measurements to be taken of depth of silt and so on. Another example is the simulation of aircraft aerodynamics. This could use a scale model of the aircraft in a wind tunnel or water tunnel, or it could even involve flying a radio-controlled model. Such simulations can provide useful data on the distribution of pressure around the aircraft.

There is another important distinction to be made in the field of modelling—that is, between static and dynamic models. By definition the characteristics of a static system do not change with time. The dynamic system, on the other hand, has time-varying characteristics. For example, a map, a two-dimensional iconic model, is clearly a static model, but the estuary model mentioned earlier is dynamic, since its characteristics would change with time as silting progresses.

6.3   Symbolic Models

Let us now move on to another category of model, the symbolic model. The symbolism can be a written language, a mathematical notation or even a thought process. Let us start with the last one.

Symbolism of Mental Images

Did you know that your perception of reality is a symbolic model of the real world? This differs from individual to individual, and the different

Fig. 6.1. A two-dimensional iconic model of the author.

interpretations we place on events, and the resultant actions that we take, are influenced by our individual perceptions or models of the real world. Thus it is not surprising that our conclusions about real events will differ. In addition, the accuracy of our mental model depends on the accuracy of the information supplied to us, and this depends not only on the real information itself, but on the ability of our senses to transmit the information accurately.

A poem by the American poet John Godfrey Saxe illustrates very nicely how our models of reality can vary when information is incomplete:

The Blind Men and the Elephant

It was six men of Indostan
To learning much inclined
Who went to see the Elephant
(Though all of them were blind),
That each by observation
Might satisfy his mind.

The First approached the Elephant,
And happening to fall
Against his broad and sturdy side,
At once began to bawl:
'God bless me! but the Elephant
Is very like a wall!'

The Second, feeling of the tusk,
Cried, 'Ho! what have we here
So very round and smooth and sharp?
To me 'tis mighty clear
This wonder of an Elephant
Is very like a spear!'

The Third approached the animal,
And happening to take
The squirming trunk within his hands,
Thus boldly up and spake:
'I see,' quoth he, 'the Elephant
Is very like a snake!'

The Fourth reached out an eager hand,
And felt about the knee.
'That this most wondrous beast is like
Is mighty plain,' quoth he:
' 'Tis clear enough the Elephant
Is very like a tree!'

The Fifth who chanced to touch the ear,
Said: 'E'en the blindest man
Can tell what this resembles most;
Deny the fact who can,
This marvel of an Elephant
Is very like a fan!'

The Sixth no sooner had begun
About the beast to grope,
Than, seizing on the swinging tail
That fell within his scope,
'I see,' quoth he, 'the Elephant
Is very like a rope!'

> And so these men of Indostan
> Disputed loud and long,
> Each in his own opinion
> Exceeding stiff and strong.
> Though each was partly in the right
> And all were in the wrong!

Going even further, some philosophers would have us believe that there is no real world, and what we perceive is all a concoction of the mind. But let us leave the hazardous field of philosophy and return to the safer pastures of technology.

Symbolism of Written Language

In addition to these mental images, there are two other important symbolisms: the **written language** and **mathematical notation.** Which symbolism is the better?

For example, consider the seesaw. We will all have different impressions or mental models of seesaws, depending on our association of the seesaw with past events, happy or otherwise. Perhaps our ideas would converge if we were to construct a symbolic model using the English language as our symbolism. The Chambers' definition is 'a plank balanced so that its ends may move up and down alternately'. But many people would define a seesaw as 'a toy found in public parks', and the Chambers' definition would not create the right image for them. Thus, unless great care is taken, symbolic models which use the English language may be unsatisfactory. A lot of effort is needed to produce a good symbolic model, so much so that we are often tempted to resort to the simpler task of making an iconic model by, in this case, flapping our arms about.

Symbolism of Mathematical Notation

Mathematical notation produces a symbolic model which is more concise, has a simpler syntax and is less ambiguous than the English language. But, of course, the **mathematical model** is a very specialised symbolic model, and is of little use to the layman. In the case of the seesaw, the mathematical model is

$$\text{Acceleration} = \ddot{\theta} = \frac{\mathrm{d}^2\theta}{\mathrm{d}t^2} = \frac{-FL_1\cos\theta}{M_2L_2(L_1 + L_2) + J} \tag{6.1}$$

where F is the applied force and J is the moment of inertia of the plank about the pivot. This equation relates the angular acceleration of the seesaw to the geometrical and mass parameters and the applied force (Fig. 6.2). It is, in mathematical jargon, a second-order non-linear differential equation, but it is just as much a model as the iconic model at the top of the diagram. It is yet another representation of reality.

However, this form of symbolic model is special—its solution predicts the time-wise variation of the system variables (see Fig. 6.2). Although Newton had for many years a good mental model of the mechanism of the

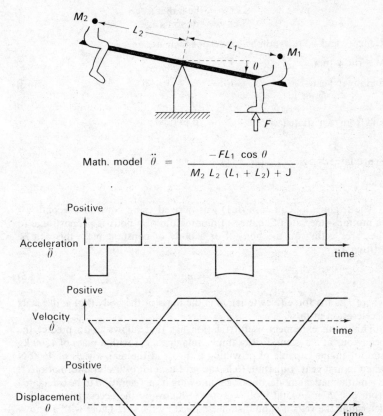

$$\text{Math. model } \ddot{\theta} = \frac{-FL_1 \cos \theta}{M_2 L_2 (L_1 + L_2) + J}$$

Fig. 6.2 Mathematical model of a seesaw. Solution of the model equations shows how the angular displacement, velocity and acceleration of the seesaw will vary with time.

universe, the full value of his genius was not appreciated until he wrote down the mathematical model which allowed eclipses and other phenomena to be accurately predicted.

Equation 6.1 is complex and the reader is not expected to understand its method of derivation or of solution. It is merely illustrative.

Fortunately, mathematical models are not always as complicated as this. For example, the mathematical model of a resistance is $R = E/I$, where E is the voltage across the resistance R and I is the current flowing. And mathematical models are not restricted to esoteric technological problems. They help us to solve simple problems such as the following: A father is three times as old as his son. In ten years his age will be twice that of his son. What is his age? The mathematical model is constructed as follows:

$$\text{Let } x = \text{the father's age}$$
$$\text{Let } y = \text{the son's age}$$

Hence x and y are symbolic models of their ages.

We know that $\qquad\qquad x = 3y$ (6.2)

and in ten years $\qquad (x + 10) = 2(y + 10)$ (6.3)

or $\qquad\qquad\qquad x - 2y = 10$ (6.4)

and from equation (6.2) $3y - 2y = 10$

or $\qquad\qquad\qquad y = 10$

hence father's age $\qquad x = 3y = 30$ years

The frightening equation (6.1) was derived using Newton's second law of motion—the rate of change of momentum of a body is proportional to the force acting on the body. For a body of constant mass this can be written as:

$$F = ma \qquad (6.5)$$

where F is the force (Newtons), m is the mass of the body (kg) and a is its acceleration in m/s^2.

This is an extremely useful relationship that allows us to predict the behaviour of a system. For example, imagine a car with a mass of 1500 kg and an engine capable of providing a thrust at the rear wheels of 4500 N when in first gear. Equation (6.5) can give us a lot of useful information. It is a mathematical model of our car allowing its acceleration to be calculated from $a = F/m = 4500/1500 = 3$ m/s.^{2}. Knowing the acceleration we can determine the velocity and distance moved by the car. Those with a knowledge of the calculus will know that

$$\text{Velocity } v = \int_0^t a \, dt = \int_0^t 3 \, dt = 3t$$

where t is the time in seconds elapsed from rest.

Hence after 10 seconds, the velocity would be $3 \times 10 = 30$ m/s.

We can also find d, the distance moved in a given time

$$\text{Distance } d = \int_0^t v \, dt = \int_0^t 3t = 1.5 \, t^2$$

Hence after 10 seconds our imaginary car would have covered a distance of 150 m.

Thus, if we can write down a mathematical model of a system, we are well on the way to predicting its behaviour. This is an invaluable design aid. If for example, the mathematical model of a proposed aircraft is available, then the dynamic characteristics of the aircraft can be assessed before it is manufactured. A lot of money and time can be saved by use of such a model during the design stage.

But can we go further and construct mathematical models of everything, including ourselves, thereby allowing us to predict our future? Not surprisingly the answer is *no*! Construction of a mathematical model of a system requires a complete understanding of the operation of the system, and that all phenomena can be described in mathematical notation. It is also necessary that all parameters can be measured accurately—for example, the aircraft's mathematical model is of little use unless accurate information relating to strength of the structure, to the engine performance, etc., is known, these all requiring painstaking laboratory investigation. In addition, it must be possible to make measurements of the variables for one cannot predict the future value of a variable if its present value is not known. Fortunately in applied science, measurement techniques are now so advanced that nearly all physical variables can be measured with great accuracy, but of course in the atomic world of the physicist, Heisenberg's principle rears its ugly head! (Heisenberg argued that it is not possible to determine both the position and velocity of an elementary particle.) Finally, a mathematical model is of little use if it is so complex that its equations cannot be solved even with the aid of computers.

The modeller is faced with a dilemma regarding the necessary complexity of the model. If it is too simple the results will be of restricted use. But as more than one modeller has realised whilst regretfully contemplating his maze of equations, it is possible to construct a model that is harder to understand than the real thing.

The modeller must fix his objectives clearly in his mind. Consider a problem concerned with the dynamic characteristics of an automobile. How complex should the model be? If we wish to study the acceleration performance it is sufficient, as we showed earlier, to treat the car as a particle acted upon by a propulsive force P and a friction force F (Fig. 6.3). The model is then

$$M\ddot{x} = P - F \tag{6.6}$$

where $\ddot{x}$ is the acceleration of the car (the a in equation 6.5).

However, if we wish to ensure that the passenger has a comfortable journey in the vehicle, the effectiveness of the suspension must be investigated. This can be achieved by combining all of the spring elements, and by combining all of the energy-dissipating elements—in this case the shock absorbers (Fig. 6.4). This is known as 'lumping' the parameters. In addition, the road profile is now of great importance since it provides the input excitation for the system. For example, if the car ran over a brick the input excitation Y could be represented by the upper graph of Fig. 6.4. The response of the car, Xc, could be calculated and might look something like the lower graph.

The model grows in complexity if the forces on the stub axles are to be examined, since the dynamic characteristics of the wheels themselves now need to be identified (Fig. 6.5). Again, even further complexity is necessary when the effect of road profile on braking characteristics is to be evaluated.

This process of growing complexity can be continued indefinitely, but the important thing is to know where to stop. It is even more important to

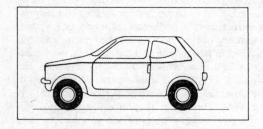

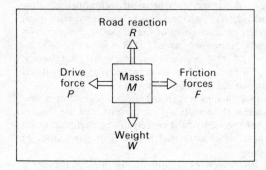

Math. model

$$M\ddot{x} = P - F$$

Fig. 6.3. The simplest mathematical model of a car.

remember where you stopped! Many instances can be quoted where assumptions used during model construction were ignored during later application of the model. One common assumption, which leads to a **linear model** whose equations can be solved analytically, is that of small perturbations. Referring back to equation 6.1 you will note the presence of a $\cos\theta$ term. This is a **non-linearity**, for the obvious reason that $\cos\theta$ does not vary in a linear manner with θ, i.e. equal increments in θ do not give equal changes in $\cos\theta$. Now non-linearities make equations like 6.1 very difficult, if not impossible, to solve analytically, so we have to find a way around this. The **method of small perturbations** helps us. If we assume that perturbations (changes) in the system variable (θ in this case) remain small during motions then we can approximate $\cos\theta$ by 1, since $\cos(0) = 1.0$. The non-linearity then disappears and the resulting linear equation is easy to solve. But we must remember that the solution is an approximation—an approximation that gets worse as θ gets larger.

The following table shows how the accuracy deteriorates as θ increases.

θ (deg)	0	5	10	15	20	25
$\cos\theta$	1.0	0.99619	0.98481	0.96592	0.93969	0.90631

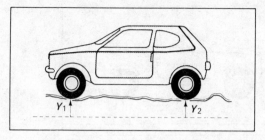

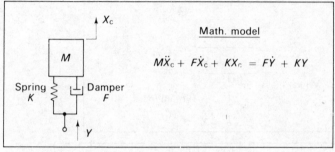

Math. model

$$M\ddot{X}_c + F\dot{X}_c + KX_c = F\dot{Y} + KY$$

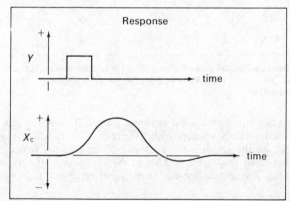

Fig. 6.4. A more complicated mathematical model of a car, allowing a study of the car's response X_c to the up-and-down motion of the road surface Y.

This must be borne in mind and you must be careful not to apply the equation for large θ, for the error will be unacceptable.

6.4 Analogue Models

We have now met iconic models and symbolic models, and the last main category we have space to discuss is that of analogue models. Analogues use one set of properties to represent another set of properties. For ex-

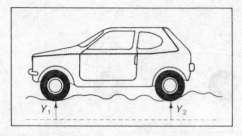

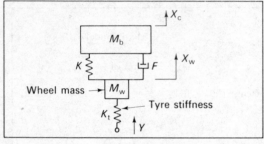

Math. model

$$M_w \ddot{X}_w + F \dot{X}_w + (K + K_t) X_w = K_t Y + K X_c + F \dot{X}_c$$

$$M_b \ddot{X}_c + F \dot{X}_c + K X_c = K Y + F \dot{Y}$$

Fig. 6.5 A mathematical model of a car that allows for relative motion between the chassis X_c and the wheels X_w in response to the up-and-down motion Y of the road surface.

ample, a graph is a two-dimensional analogue model. It may be a plot of the variation of a rocket's velocity with time (Fig. 6.6), in which case a vertical distance on the graph is the analogue of the rocket's velocity; it varies with time in an analogous manner to the variation of the rocket's velocity with time. The vertical distance on the graph represents the velocity of the rocket.

Another useful set of analogues can be derived from systems having analogous mathematical models. For example, can you spot the similarity of the mathematical models of the mechanical and electrical systems in Fig. 6.7? They are analogous, and this allows us, for example, to use the variation of current in the second one as a representation of the variation of velocity in the first or of voltage in the third.

Analogue models are very useful. In the case shown it is possible to assess the effect of changes in mass in the mechanical system simply by turning a knob to vary the inductance in the electrical system. In addition, one does not have to use mathematical techniques to solve the equations.

There are many interesting analogues. For example, you may find it hard to believe that we can glean information about stress and strain in a

elocity

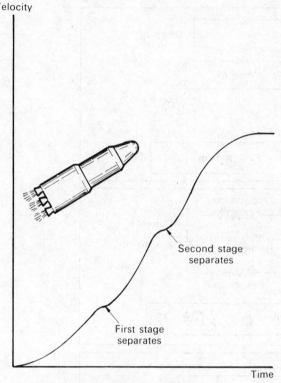

Second stage
separates

First stage
separates

Time

Fig. 6.6. A graph is an analogue model. Here the vertical distance is an analogue of
the rocket's velocity.

mechanical system by studying soap bubbles; and we can draw conclusions
about traffic flow between cities by studying the gravitational attraction
between bodies. There are electrical analogues of the pulsating flow of
liquid in a pipe, and of the stresses set up when steel rods are twisted.

6.5 Computers in Simulation

Analogue Computers

We referred briefly to the analogue computer in Chapter 5. It is a very
popular **simulator.** It employs a number of elements, each used to represent
a mathematical operation such as addition, integration, multiplication.
These elements are interconnected in a manner determined by the mathematical model of the system to be simulated. Fig. 6.8 shows a simplified
mathematical model of an aircraft and the corresponding analogue computer circuit, so arranged that the dynamic behaviour of circuit voltages is
described by similar mathematical equations as the aircraft variables.

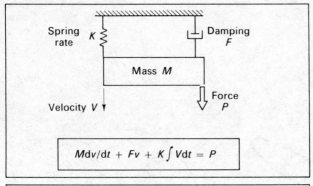

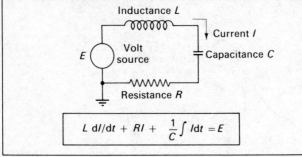

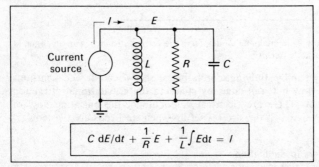

Fig. 6.7. Mechanical and electrical systems with analogous mathematical models.

Voltages in the computer therefore represent the actual system variables. A big advantage is that the time scale can be varied, allowing the mathematical models of fast phenomena such as those found in electronic circuits to be slowed down, and allowing slow phenomena such as the thermodynamics of large structures to be speeded up.

Analogue flight simulators are extensively used in the aircraft industry, both during the design phase and for pilot training. First a mathematical model of the aircraft is established. An analogue computer is then used as a model so that circuit voltages behave in a similar manner to the aircraft

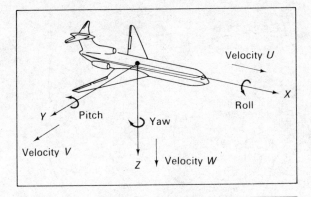

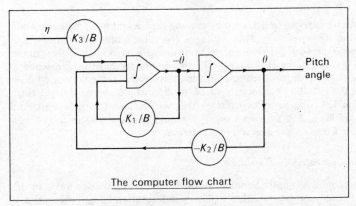

Fig. 6.8. An analogue computer circuit to study the pitch of an aircraft. The circuit includes two integrating amplifiers and three potentiometers.

variables such as speed, pitch angle, etc. This simulated aircraft can now be 'flown' by applying control voltages instead of the actual stick and throttle movements required in the actual aircraft (Fig. 6.9). To give the simulation realism, the pilot can be placed in a mock-up of an aircraft cockpit so arranged that movements of the controls generate voltages which act as inputs to the analogue computer. The outputs from the computer can then be used to drive the cockpit instruments such as the airspeed indicator so that the pilot gets the impression of flying the actual aircraft.

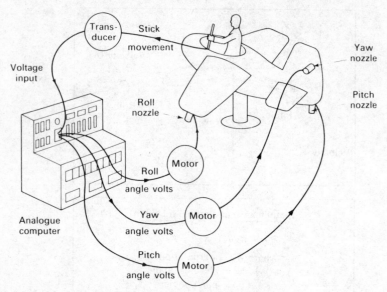

Fig. 6.9. An aircraft simulator using an analogue computer.

Greater realism can be achieved by making the cockpit move in accordance with the appropriate output voltages from the computer. Since these voltages represent the real aircraft's motion, the cockpit will then move about as would the real aircraft. Flight simulators have now reached a very advanced stage of realism. In the design field they enable a pilot to assess the handling characteristics of an aircraft which has not yet been built. When used for pilot training, they save a lot of expensive time in flight training. A crash of a Concorde would be disastrous in reality, but simulated crashes are everyday occurrences.

Digital Computers in Simulation

The analogue computer handles information in a continuous form. In the digital computer, on the other hand, the information being processed can only assume a finite number of discrete states. As we have seen in Chapter 5, digital computers use binary arithmetic for their calculations, thus involving only two digits 1 and 0. In *Alice in Wonderland*, the White Queen asked Alice: 'Can you do addition? What's one and one and one and one and one and one and one and one and one and one and one?' 'I don't know,' said Alice, 'I lost count.' Digital computers reduce all of their calculations to this form but fortunately they don't lose count. As far as dynamic modelling is concerned, analogue computers are faster, but digital computers are more accurate and can store data. There are many applications where the nature of the problem demands digital computing—for example, in accountancy, where many repetitive arithmetic operations are involved. And in simulation, if the solution of the model's behaviour involves decision-taking stages, the digital computer is needed.

Consider the problem of routing a vehicle through a network of roads or guideways so that it will always get from one point to another in minimum time. This is now an important consideration for one of the proposed new methods of transport in cities—the autotaxi or the personal rapid transit system. A plan, or iconic model, of a network is shown in Fig. 6.10. There are 16 stations and the total route length in 85 km. One-way traffic is assumed, requiring the network and intersections shown at the bottom of the figure. Now the objective is to minimise travel time, so it is necessary

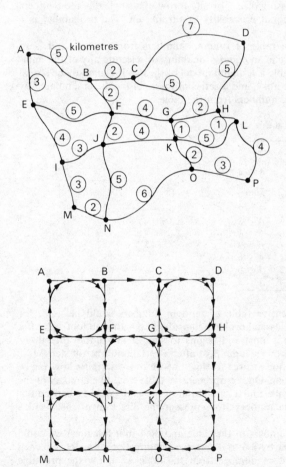

Fig. 6.10. A traffic routing network—one-way traffic only.

to determine the quickest route from any point on the network to any other point. The digital computer is ideal for carrying out this type of calculation, and its use would be essential for large networks, for even in this simple case there is a total of 256 optimum paths among stations. Proposed systems would use a digital simulator to direct autotaxis on to these optimum routes, and would continually update this information, taking into account heavy loading on some routes and breakdowns, etc. We shall refer to this system again later in this chapter.

Digital, as opposed to analogue, simulation is also widely used for the study of **non-deterministic systems.** These are sometimes called probabilistic systems and, as the name implies, in such systems some of the relationships are not uniquely defined—we cannot say exactly what value a particular variable will have, but we know what it will probably be. A very simple example is that of tossing dice. For any throw of a single die, the numbers 1 to 6 all have an equal **probability** of turning up. The probability is 1 in 6.

Table 6.1 shows a table of random numbers from 1 to 6, and is so constructed that there is an equal probability of selecting any of the numbers 1 to 6. So this table is a digital model of the die. It can easily be stored in the computer's memory, and a series of tosses of the die is simulated by the computer picking numbers from the table.

Table 6.1.   Model of a die

2	5	6	5	4	5	5	4
2	4	2	6	2	4	5	3
4	1	5	6	2	1	4	4
1	3	5	3	6	6	5	6
6	4	2	3	5	4	3	3
5	5	2	1	6	1	3	2
6	2	4	2	2	3	4	3
6	2	4	1	1	6	5	1
1	1	3	5	3	1	6	2
3	5	1	6	1	1	4	6
2	4	1	3	2	3	3	6
4	5	2	6	3	4	1	5

A more comprehensive table of random numbers could be used, for example, to study a classical problem in probability—the random walk. If a drunk starts at a lamp post and begins to stagger in random directions, how far will he be from the lamp post after a certain number of steps? It is clear there is no unique answer to this, but we can determine how far he *probably* will have gone. Our comprehensive table would be a model of the randomness of his walk, and a simulation of his walk could be carried out by using the random numbers to represent the direction of the drunk's steps.

If there are 100 numbers in the table and 360° in a full turn, we could multiply each number by 3.6 to determine the direction of a step. So each time we take a random number from the table we know the probable direction of the drunk's next step. Fig. 6.11 shows his progress after 75

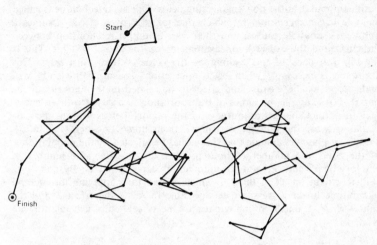

Fig. 6.11. The drunk's walk—an example of digital simulation.

steps. We could repeat this simulation many times and by compiling all of the results we would determine the probability of the drunk walking beyond a certain point.

This is an amusing problem, but there are many more practical problems that require a similar technique. Queuing problems are of this nature—whether it be queues of cars at traffic lights or queues of customers at a counter. When we do not know precisely how many cars or people there are, or when they will arrive or leave, we have to use a non-deterministic model.

6.6 Empirical Models

Modelling and simulation are important tools for the technologist, but they are also finding application in other fields such as economics, town planning, marketing and sociology, where it is usually difficult, if not impossible, to write down exact mathematical relationships between variables. This is easier in science and technology, where a multitude of basic laws exist such as Ohm's Law, Charles' Law, Hooke's Law and Newton's Laws, and these, used in combination with experimentally determined properties of components and systems, allow symbolic mathematical models to be formulated even for the most complex engineering systems. You can, for example, write down an equation relating current and voltage in an electrical circuit, but you would have a problem trying to write an equation for the relationship between the cost of petrol and the number of people travelling by bus. Nevertheless there is a relationship, because an increase in the cost of petrol can persuade people to leave their cars at home and take the bus instead.

Because of this difficulty, so-called **empirical model** techniques have become popular in the non-engineering fields. One such technique is called **multiple linear regression**, by which one takes a given effect, assumes a number of possible causes, and then tries to find a relationship between the effect and the causes by measurements on the actual systems. This is the big drawback of this technique—the actual system must exist. This approach is necessary when basic interactions are not sufficiently well understood to allow cause and effect to be related mathematically at the outset. For example in studies of transportation in a city it is often necessary to predict the future journeys in the region. Let us look at the particular case of the number of trips T from home to work (Fig. 6.12). Transport planners assume that this would be affected by the population P of the zone, the number of households H in the zone, the number of employed residents E, and the number of cars owned C. By collecting sufficient data on all of these parameters, and then applying the method of multiple linear regression, it is possible to relate the effect T and the causes P, H, E and C by an equation. One typical empirical relationship is

$$T = 0.097P - 0.351H + O.773E + 0.504C - 43.6 \qquad (6.7)$$

But this equation only states an empirical relationship—it does not tell us why!

This equation relates the present measured values of these quantities, but such equations are often used as predictors. In this case with estimates of P, H, E and C for the year 2000, the equation could be used to predict the number of journeys from home to work in 1990. However, such empirical equations need to be treated with caution. They do not include dynamic effects—implied by the assumption that the same equation will be valid 16 years later. We all know from experience that relations can vary with time, and so a much better predictive model requires the recognition of basic laws at work in the system, thus allowing a model to be constructed without the need of data from the actual system. Of course this is easier said than done, but a lot of effort is being put into it now in the fields of social dynamics, industrial dynamics, urban dynamics, operational research, etc. Empirical models are being abandoned, and more effort is being devoted to understanding the chain of cause and effect.

6.7   Models with Feedback

Negative feedback, a clear example of cause and effect, comes into play in many systems, and when once recognised provides the starting point for a mathematical model. We met negative feedback earlier in Chapter 4, but four further examples are shown in Fig. 6.13. In the case of the tank of liquid, the outflow is determined by the level, but the outflow in turn determines the level. Feedback is negative since the outflow reduces the liquid level. Similar effects can be seen in the other systems. In fact they are all analogues described by the equation.

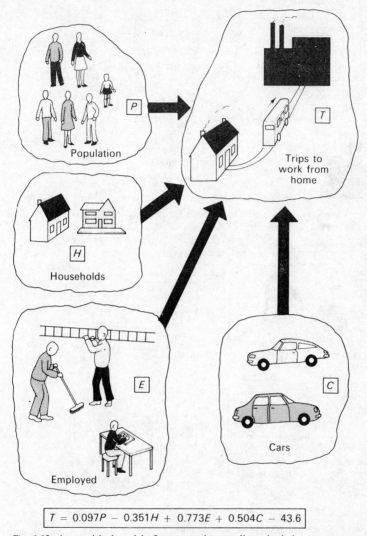

$$T = 0.097P - 0.351H + 0.773E + 0.504C - 43.6$$

Fig. 6.12. An empirical model of transport in a medium-sized city.

$$X + T\frac{\mathrm{d}x}{\mathrm{d}t} = 0 \qquad (6.8)$$

where T is positive and X is the appropriate variable of level, capital, age or population. (Since equation (6.8) is a first-order differential equation these systems are often referred to as first-order systems.)

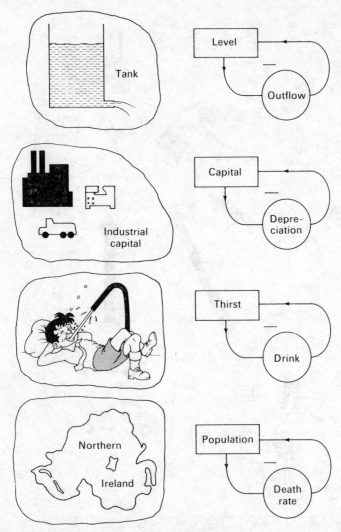

Fig. 6.13. Systems employing negative feedback.

The accuracy of this mathematical model will vary from system to system, probably being most accurate for the tank example, and least accurate for the thirst satiation example. But this equation can be solved so that we are able to predict the behaviour of such systems. They all tend to settle out to a steady state, the time required increasing with T which is often referred to as the time constant.

Positive feedback can also be found in many systems. Fig. 6.14 shows four more examples to be added to the microphone speaker system of

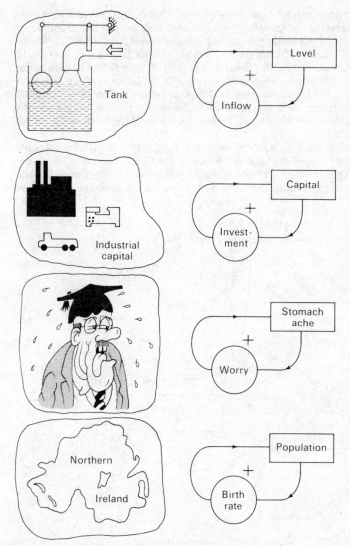

Fig. 6.14. Systems employing positive feedback.

Chapter 4. In these cases runaway situation arises, increases in output level leading to increases in input which in turn lead to increases in output. The mathematical model for this family of analogue systems is

$$X - T\frac{\mathrm{d}x}{\mathrm{d}t} = 0 \qquad (6.9)$$

Solution of this equation indicates, as expected, a runaway response.

The responses exhibit **exponential growth**—that is, they increase by a constant percentage of the whole in a constant time period. For example, let us pursue the population model, but consider yeast cells instead of Northern Irish citizens. If each cell in a colony divides into two cells every 10 minutes, there will be an exponential growth of yeast cells. For each single cell, after 10 minutes there will be two cells, an increase of 100 per cent. After the next 10 minutes there will be four cells, a further increase of 100 per cent. Hence the population will increase by a further 100 per cent (it will double) every 10 minutes. In Northern Ireland the birth rate is not so dramatic.

In reality these runaway situations are limited by physical constraints such as the size of the tank or the size of the Earth. The explosive nature of the exponential growth of population had a startling effect on the public in 1790, when Malthus published his famous work *A Summary View of the*

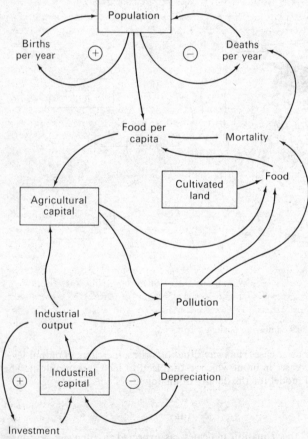

Fig. 6.15. Some of the feedback loops in the world model.

Principle of Population, in which he predicted that population growing exponentially would very soon exhaust food supplies since arable land only increased arithmetically. (In arithmetic growth, a constant amount is added over each time interval.) In other words, food production must always lag behind birthrate! Malthus's predictions were inaccurate, because his model neglected several important interactions. But modelling is now more of a science than an art, and already the new methods of system dynamic modelling, pioneered by Forrester and Massachusetts Institute of Technology (MIT), are being used to good effect. Forrester's model of an urban area, for example, evoked considerable discussion, and indeed, if only for that reason, has made a noteworthy contribution to the understanding of the urban area. Modelling of social systems will always lead to controversy, basically because it requires the quantification of value relationships which by their very nature differ from individual to individual.

Perhaps of more interest to us all is the dynamic model of the world, studied by the Club of Rome, MIT, and more recently by the University of Sussex. The five main model variables are population, capital, food, pollution and non-renewable resources. Fig. 6.15 shows one set of feedback loops connecting population, capital, agriculture and pollution. The two basic positive feedback loops of population and capital are prominent. These interact; some output from capital is diverted to agriculture (tractors, fertilizers, etc.); this in turn affects the amount of food produced, and again in turn affects mortality. This is a complicated set of interactions, but it is only a small part of the dynamic model. Each of the arrows on the flow diagram represents a causal relationship, and for computer solution it is necessary to quantify these relationships. As expected, this is the part of the model that has received most criticism, but the results are intriguing. They predict (see Fig. 6.16) that if things continue as they are—i.e. if there are no changes in the physical, economic or social relationships that have to date affected the world's development—then we will be in trouble in the near future. Food, industrial output and population will grow exponentially until growth is stopped by reduction in resources. Population begins to fall off about AD 2050 because of reduced medical services and food shortages. This is a grim picture, and indeed has caused quite a stir.

The world model is certainly on a grand scale. On a lesser scale, at the national level, economic models are now being used as aids to policy-making. The complexity of today's economy does not allow the government to rely on common sense or inspired guesswork for policy formation. So the British Treasury now has a model of the national economy at its disposal.

There is no doubt that the use of computer simulation can give an insight into the operation of systems which include many interactions and which behave in a counter-intuitive manner. But there is a danger in ascribing to the model output a sanctity which it does not possess. The value of the results depends on the accuracy of the model. It is unfortunate that computer results are often used to support arguments by people who do not understand the basic assumptions and the shortcomings of the model. On the other hand, the results, especially when controversial, like

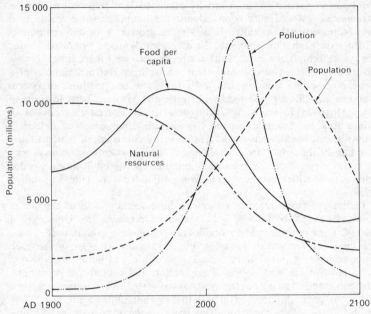

Fig. 6.16. Some predictions from the world model.

those of the world model, are often attacked for reasons other than logical ones.

6.8   Optimisation

When a technologist seeks to optimise an artifact or technique, he strives to make it as good as possible with respect to some **criterion**. The criterion could be performance, efficiency, strength, appearance and so on, and indeed any combination of these. We can quote many examples of optimisation from our own everyday experience. When we settle down in front of the fire on a winter's evening we attempt to find the optimum distance from the fire at which our comfort, the criterion in this case, is a maximum. When we sweeten our tea there is an optimum amount of sugar that gives our palates the maximum sense of pleasure. And we can even recognise optimisation in action when we consider the aesthetic pleasure derived from our appreciation of shape and form. Look at Fig. 6.17 and choose the shape that you consider to be most pleasing. Many people would agree that the rectangle (g) with proportions 1 to 1.618 is the most beautiful. This shape is known as the golden section or golden ratio, and it has long been used by artists, such as Titian and Michelangelo, to enhance the attractiveness of their work. So there is an optimum shape that gives us the greatest amount of aesthetic pleasure.

In the above cases we wish to maximise desirable properties like comfort

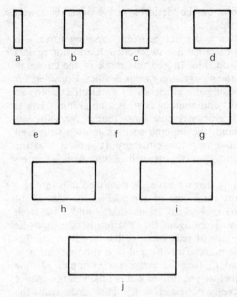

Fig. 6.17. Choose the rectangle that pleases your eye (see text).

and pleasure, but there will be other cases where the optimum occurs when some undesirable aspect is at a minimum. For example, if you have a choice of several routes between two towns you will normally choose the shortest route since time spent on travelling is time wasted and is therefore undesirable. Then again, when buying the groceries we would wish to minimise cost. Thus, in general, optimisation involves minimising the undesirable and maximising the desirable.

In many optimisation problems the relationship between the criterion (the dependent variable) and the system parameter (the independent variable) looks like that shown in Fig. 6.18. The technologist strives to operate

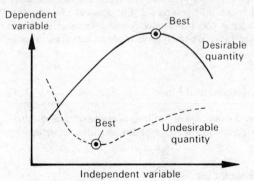

Fig. 6.18. Optimisation—getting the most of a desirable quality or the least of an undesirable one.

the systems at the topmost point on the 'desirable' curve or at the lowest point on the 'undesirable' curve.

Optimisation usually involves a conflict between opposing forces—in our earlier examples, too sweet against not sweet enough, too hot against too cold, and too fat against too thin. Indeed, the shape of the curves in Fig. 6.18 can be explained by the recognition of this conflict. Consider, for example, the problem of choosing the efficiency of an electric motor so that its overall cost (motor cost and running cost) is a minimum. Here is the conflict: motors with high efficiency cost more than those with low efficiency but, on the other hand, the running costs of motors with high efficiency will be less than those with low efficiency. It is clear that the 'best' motor, that with the least overall cost, will be one with an intermediate efficiency.

The very business of modelling that we have just described falls into this category too. Let us define the total cost of a model as the cost C_d of developing the model plus the cost C_e of errors that would arise from applying the model in practice. Here again the criterion is cost, and the independent variable is the degree of refinement of the model. A conflict arises between C_d and C_e. The more effort we put into the refinement of the model, the greater the cost C_d (see the earlier car example). On the other hand, the greater the effort we put into modelling the less the chance of the real system having a costly design error C_e. Thus once again the optimum lies somewhere between the extremes.

And, familiar to all of us, think of a toothbrush bristle. If it is not stiff enough it will not clean our teeth properly, but if it is too stiff it will make our gums bleed. There is an obvious conflict here, but in this case the criterion is more complicated than in our other examples where cost dominated. Here we want a toothbrush that is comfortable and effective. Thus our criterion must reflect this and, for example, we might wish to optimise the sum of effectiveness and comfort. This is not so easy and we shall examine it more closely in Chapter 7. Indeed, the criteria upon which we base our optimisation can become very complicated, and this is illustrated by the simple business of buying a car. A purchaser would be looking at cost, speed, comfort, reliability, appearance and so on, and the chosen car, his 'best', would be the one that provided the right mix of these attributes. By the right mix I mean that *you* might give comfort priority over speed, whilst *I* might prefer it the other way round. So value decisions come into this optimisation task. We shall have more to say about this in the next chapter.

6.9   Optimisation using Mathematical Models

We shall now look at several examples of optimisation where mathematical models can be developed, allowing the use of the calculus to determine maxima and minima.

Designing a Tray for Minimum Cost

Let us assume that we wish to make trays from a stock of sheets of plastic.

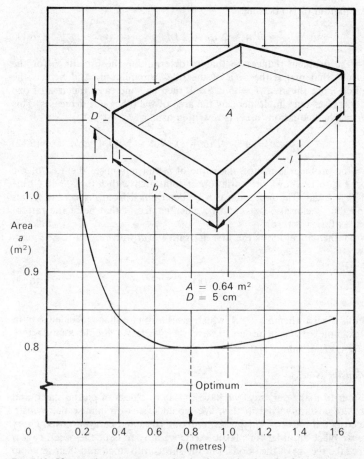

Fig. 6.19. If a tray has an area *A* and a depth *D*, what length of side minimises the amount of materials required? The area is a minimum when the tray is square; here *b* = 0.8 m.

We know the volume *V* of liquid that the trays must hold, and we are told that the trays must have a depth of *D* (Fig. 6.19). (Throughout these examples we shall use capital letters for fixed quantities.) Our problem is to produce a tray that uses the minimum amount of our plastic sheet. Let us see if we can identify the conflict in this case. If the length of the tray is *l* and the breadth *b*, then the total area of plastic needed is

$$a = \text{area of base} + \text{area of sides}$$

or
$$a = bl + 2Db + 2Dl \qquad (6.10)$$

Now the area of *bl* is fixed, since we are given both *V* and *D*. Hence

$$bl = V/D = A$$

and equation (6.10) becomes

$$a = A + 2Db + 2Dl \qquad (6.11)$$

So the problem reduces to that of determining the breadth b (or the length l) that makes the area of the sides a minimum. And here is the conflict. Since the area $A = lb$ is fixed, then as b increases the area of one side increases but simultaneously the area of the other side decreases. This is clear when equation (6.11) is rewritten as

$$a = A + 2D(b + 1) = A + 2D(b + A/b) \qquad (6.12)$$

So the problem is to find the value of b that minimises the area of the sides. Fig. 6.19 shows the variation of total area with b for $A = 0.64$ m^2 and $D = 5$ cm. The minimum can be determined from inspection of this graph, but it can also be found by calculus, for at that point the rate of change of area is zero.

In mathematical terms the **first derivative** of a with respect to b is zero. Now

$$\frac{da}{db} = 2D(1 - A/b^2) \qquad (6.13)$$

and this is zero when $b = \sqrt{A}$, which gives a square shape (of side 0.8 m in the example). Thus in order to save on plastic we should manufacture square trays of side $\sqrt{V/D}$.

Optimising Stocks

In order to make our trays we have to buy in stocks of plastic sheet, and there are costs involved in this. We would wish to minimise these costs. The costs fall into two categories: **ordering costs** and **carrying costs**. Every time we place an order for plastic sheets we have to issue the order, follow it up, take receipt of the goods, put the sheets into stock and then settle up with the manufacturers of the sheets. These ordering costs are much the same no matter what size the order may be. So we could keep down our ordering costs over the year by making only a few orders for large quantities instead of a lot of orders for small quantities.

> Let N = total number of sheets required per annum
> n = size of order
> P = cost of making a single order

Thus, annual ordering cost

$$c_1 = PN/n \qquad (6.14)$$

But on the other hand, if we order a large number of sheets, our stock-holding costs will be high, basically because we are locking up money capital in goods. This can be explained with the help of Fig. 6.20(*a*). Assuming that we need 12 000 sheets per annum, then we could, for ex-

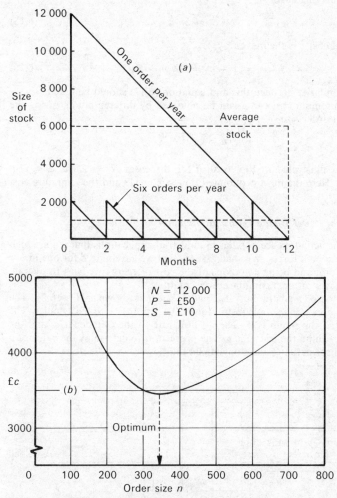

Fig. 6.20. Optimising stock: (*a*) the effect of placing one big order or six smaller ones; (*b*) the economic order size is 347 items.

ample, place one order for 12 000 sheets or six orders for 2000 sheets each. If the sheets are used at a uniform rate you can see that the average number of sheets in stock over the year is one half of the order size. Holding this amount of material in stock costs money because we have the alternative of investing the money in a bank or a project where it would accrue interest. Thus, for this reason, we would seek to hold as little as possible in stock by reordering small quantities at regular intervals. This conflicts with the conclusion derived from our discussion of ordering costs.

If S equals the cost of holding one unit in stock for one year, then the annual carrying cost

$$c_2 = Sn/2 \tag{6.15}$$

The total cost c is then given by

$$c = c_1 + c_2 = PN/n + Sn/2 \tag{6.16}$$

The similarity between this and equation (6.12) should be noted.

The minimum cost can again be obtained by differentiation, giving the optimum order size

$$n = \sqrt{2PN/S}$$

This is illustrated in Fig. 6.20(b) for the cases $N = 12\,000$, $S = £10$, $p = £50$. Here the most economic order size is 347 and the minimum cost is £3464.

Maximum Power from a Hydraulic Actuator

Hydraulic actuators are popular in industrial applications that require high power and high forces. We shall assume that we have a need for one in our tray factory, and being conscious of the cost of energy we wish to operate our actuator at its maximum power condition.

A sketch of a hydraulic valve and cylinder is shown in Fig. 6.21. The valve is shown as a simple restriction in the supply line to the cylinder. The pressure at the pump is P. The oil flow rate to the cylinder is q, and the pressure behind the piston is p. The problem is to determine the value of p at which the actuator develops maximum power.

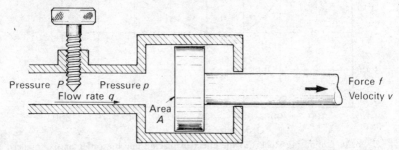

Fig. 6.21. An hydraulic valve and cylinder. Power will be a maximum when the pressure p in the cylinder is two thirds of the supply pressure P.

Our mathematical model is as follows:

The fluid mechanics of turbulent flow through orifices gives us

$$\text{Flow through restriction } q = k\sqrt{P - p} \tag{6.17}$$

where k is a constant depending on the size of the orifice and on the density of the oil.

If the cross-sectional area of the piston is A, then

$$\text{Force exerted by piston } f = pA \qquad (6.18)$$

If we assume that the oil is incompressible, then

$$\text{Velocity of piston } v = q/A \qquad (6.19)$$

Now, power is the rate of doing work, or force times velocity; hence

$$\text{Power developed by piston } h = fv = pq \qquad (6.20)$$

Now we can see the conflict, for as p increases the flow rate q will decrease because the pressure drop over the valve has decreased (equation 6.17). We have

$$h = kp\sqrt{P - p} \qquad (6.21)$$

and the reader may wish to plot h against p to convince himself that we do indeed get the curve typified by Fig. 6.18. Differentiation shows that maximum power is achieved when $p = 2P/3$. Therefore we would get the best from our actuator if it drives against a force of $2PA/3$ (equation 6.18), with a velocity of $(k\sqrt{P/3})/A$ (equation 6.19).

Minimising the Amount of Material in a Structure

Engineers have to produce many different types of complicated structures, including aircraft, bridges and buildings. Structural materials are expensive and the engineer is always striving to come up with the cheapest structure that is capable of carrying the given loads. This will usually be that structure requiring the minimum amount of material.

Consider the simple **pin-jointed truss** of Fig. 6.22 (a pin-joint allows rotational movements between the parts). Let us assume that it is part of the supporting structure of the hydraulic actuator we have just discussed. The force F is developed by the hydraulic cylinder and the geometry of the machine fixes the spacing of the feet of the truss at $2D$. We wish to determine the height h that minimises the amount of material in the structure. Each leg is of length l.

The mathematical model will be developed with reference to Fig. 6.22. Since pin-joints are used, the forces must lie along the structural members. Neglecting the weight of the structure and resolving forces along the member, gives

$$f\sin\theta = F/2 \qquad (6.22)$$
or $\qquad\qquad f = F/(2\sin\theta) = Fl/2h \qquad (6.23)$

Let S be the maximum safe stress, or force per unit cross-sectional area, that we are prepared to subject the material to. Then the minimum cross-sectional area a of the member is given by

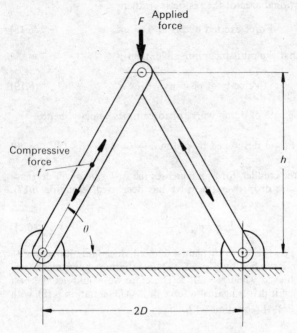

Fig. 6.22. Minimising the material in a simple structure.

$$a = f/S = Fl/2hs = \frac{F\sqrt{h^2 + d^2}}{2hS} \tag{6.24}$$

The volume v of material in the structure is given by

$$v = 2la = 2a\sqrt{h^2 + d^2} \tag{6.25}$$

Now we can spot the conflict. As h increases the member gets longer, and we would thus expect the volume of material to increase. But against this, equation (6.23) tells us that the force f in the member will decrease as h increases, and equation (6.24) will allow us to reduce the cross-sectional area of the member thereby to reduce its volume.

Combining equations (6.24) and (6.25) gives

$$v = \frac{F}{S}\left[\frac{h^2 + D^2}{h}\right]$$

or $$v = \frac{F}{S}(h + D^2/h) \tag{6.26}$$

and again the reader should compare this with equation (6.12).

Differentiation of equation (6.26) with respect to h shows that the volume is a minimum when $h = D$, or when $\theta = 45°$.

Minimising Distance

Again, from an economic point of view, it is often necessary to minimise the distance between points on a machine, building or system. There is a lot of electrical wiring in a house or in an aircraft, and it would be desirable to keep it to a minimum. Similarly, the transport engineer would wish to keep to a minimum the total length of rail required to connect a number of destinations. We shall continue our quest of optimising the tray factory, and the problem on hand is illustrated in Fig. 6.23.

Two machines A and B need a supply of compressed air. The compressed air supply enters at a point on the wall of the factory. If the distances D, N and Q are determined from considerations of safety and manoeuvrability, what should the distance x be in order to minimise the total length of pneumatic line (and reduce pressure losses to a minimum)?

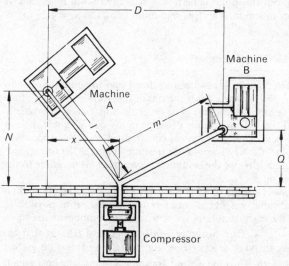

Fig. 6.23. Minimising the length of pneumatic hose, hence its cost and its pressure loss, to two machines.

The conflict is fairly obvious here. When x is small the length of the line to machine A will be small, but that to machine B will be large. For large x the reverse is true. We have

$$l + m = \sqrt{N^2 + x^2} + \sqrt{Q^2 + (D - x)^2} \qquad (6.27)$$

Differentiation with respect to x and equating to zero gives

$$x\sqrt{Q^2 + (D - x)^2} = (D - x)\sqrt{N^2 + x^2} \qquad (6.28)$$

Squaring both sides
$$x^2Q^2 + x^2(D - x)^2 = (D - x)^2N^2 + (D - x)^2x^2$$
or
$$x^2Q^2 = (D - x)^2N^2$$
or
$$xQ = (D - x)N \quad \text{or } (x - D)N \quad (6.29)$$

since a number has both positive and negative square roots.

Hence, the derivative of $d(l + m)dx$ will be zero when
$$x = DN/(N + Q) \quad \text{or} \quad x = DN/(N - Q) \quad (6.30)$$

The reason for the occurrence of two values of x will be clear if the student draws a graph of $(l + m)$ against x (equation 6.27). Taking, for example, $D = 10$ m, $N = 5$ m and $Q = 3$ m, you will find that the graph shows a minimum at $x = 6.25$ m (the first value of x in equation (6.30)). But something odd happens at $x = 25$ m (the second value of x in equation (6.30)). Here the curve momentarily flattens; the derivative $d(l + m)/dx$ is zero at $x = 25$ m, but this point is neither a maximum nor a minimum. It is called a point of inflexion. The true minimum occurs at $x = 6.25$ m.

6.10   Can we Achieve the Optimum?

In all of the above examples it has been assumed that the optimum condition is practicable. But in many cases the physical **constraints** do not allow us to achieve the desired optimum. Let us recall the sugar and tea problem, and assume that I like two spoonfuls of sugar in my cup. What if there is only one spoonful left? This is a physical constraint. I want two spoonfuls, but there is only one. Clearly I would not hesitate to accept the one spoonful because, although it is not the optimum, nevertheless it is nearer to the optimum than having no sugar.

The same argument applies to all of the other examples: if the optimum is not physically realisable, we should endeavour to get as near as possible to it. Look again at the manufacture of trays, where the optimum shape was a square of side $b = \sqrt{A}$. Our customer comes along and tells us that the trays have to pass through a letter box and therefore cannot be greater than $0.7\sqrt{A}$ in breadth. Thus the nearest we can get to the true optimum is to put $b = 0.7\sqrt{A}$.

Again reverting to the stock-holding problem, we saw that the most economic order size was 347 sheets; but if our storage area were small we may be able to accommodate only 300 sheets at maximum, and this number would be the best in the circumstances.

6.11   Operational Research Techniques for Finding the Optimum

Many optimisation problems are not amenable to solution by the calculus, and we have to resort to graphical operational research techniques. We shall first look at linear programming, a technique specifically developed to take account of the type of constraints we have just referred to.

Linear Programming

This technique of mathematical modelling is applicable to problems whose models are characterised by two kinds of linear equation. One equation relates some quantity, such as cost or weight, to the system's parameters. And there is another set of linear equations describing the constraints on the operation of the system. Let us look at an example.

A company manufactures products x and y. It makes a profit of £8 on each item x, and £10 on each item y. There are two departments A and B involved in production. Department A has sufficient staff and machines to cope with a total of 12 500 hours work per annum. In department A each item x requires 10 hours of work and each y item also requires 10 hours of work. Department B has a total annual capacity of 10 000 hours and devotes 5 hours to each x item and 10 hours to each y item. The problem is to determine what mix of items x and y should be produced in order to maximise profits.

Let us identify the linear equations. First an equation relating the profit p to the number of items x and y can be written.

$$p = 8x + 10y$$
or
$$y = -0.8x + 0.1p \qquad (6.31)$$

Next the production times for each department can be written as linear equations.

For department A: $\qquad 10x + 10y < 12\ 500$
or $\qquad\qquad\qquad y < -x + 1250 \qquad (6.32)$

For department B: $\qquad 5x + 10y < 10\ 000$
or $\qquad\qquad\qquad y < -0.5x + 1000 \qquad (6.33)$

Note the use of the inequalities.

A graphical technique is employed (see Fig. 6.24). First on a graph of y against x, we draw the straight lines (equations (6.32) and (6.33)). repre-

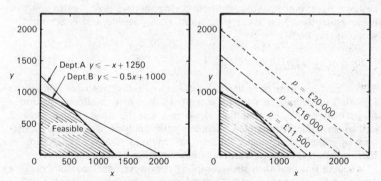

Fig. 6.24. Linear programming—determining the mix of components x and y to produce maximum profit. Here profit is maximum at £11 500 where 750 of y and 500 of x are manufactured.

senting the maximum capacity of the departments A and B. These are the constraints. Department A, for example, could produce 1250 of y and nothing else, or 1250 of x and nothing else. Intermediate amounts, such as 500 of y and 750 of x, could also be produced and all of these intermediate amounts lie on the straight line for department A. Note that this line is the maximum output from department A. Points lying under the line represent quantities that can be made without fully using the capacity; points on the line use up all the capacity; but points above the line exceed the capacity of the department. Thus the shaded area is the **area of feasible solutions**—the area where both departments can cope. The corner point occurs at $x = 500$, $y = 750$.

We now have to determine the maximum profit, so we must use equation (6.31). If we assume a profit p of £20 000, this equation becomes

$$y = -0.8x + 2000$$

and this straight line can be drawn as shown. It does not intersect the area of feasible solutions, therefore a profit of £20,000 cannot be achieved. But by continuing to draw lines of constant profit we can discover that the one for $p = £11\,500$ intersects the corner of the area and this then is the maximum profit, achieved by making 500 of x and 750 of y.

This is a very simple example involving only two constraints. The power of the method becomes more evident when there are many constraints. In general, it is found that the optimum point lies at one of the corners of the area of feasible solutions.

The astute reader will have spotted the basic difference between this example of optimisation and the earlier ones. In the tray problem, for example, the criterion was the total area of material, and this changed in a non-linear manner as the length of the side was varied. This gave the curve typified by Fig. 6.18, with its obvious maximum or minimum. But in linear programming we are concerned with linear systems in which the relationship between the criterion and the system parameters can be drawn as a *straight* line. Clearly there can be no maxima like Fig. 6.18 on a straight line, and an optimum solution can only exist if there are constraints on the design space. As we have seen, these constraints are described by a set of linear equations. Non-linear programming is possible, but is beyond the scope of this book.

Critical Path Analysis

This technique examines the subdivision of a task into the individual jobs that contribute to its achievement. It allows us to allocate priorities to particular jobs, so that the whole task can be done in the quickest and most effective manner. Critical path analysis is of great assistance in complicated projects like aircraft manufacture, the completion of housing estates, the introduction of new taxation systems and so on.

We shall not deal with such complexity, however. Instead, let's make a cup of coffee. Have you ever thought closely about this difficult task? It involves filling the kettle, plugging it in, finding the cup, the coffee,

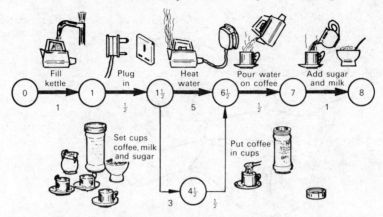

Fig. 6.25. Critical path analysis (CPA). The critical path is shown by the heavy arrows. Shortening the time of any activity on the critical path will reduce the overall time to make the coffee.

the milk and the sugar, putting the coffee in the cup, pouring the hot water, adding the sugar and the milk. Critical path analysis helps us to unravel this apparently tangled web of jobs. First we have to identify the various jobs or activities, and then we have to measure or estimate the time required to complete each of them. Using arrows and circles, it is then possible to build up a network of all the activities necessary to make a few cups of coffee (Fig. 6.25). The arrows represent the activities and the circles mark the beginning and end of each activity.

If we add up the times we can find out how long the project will take from start to finish. You can see that there are two paths through this particular network. The path that requires the longer time is known as the **critical path**, and it is identified in this example by the use of a thick line. Can you see why this is called the critical path? All activities on this path are critical in the sense that changes in their time of completion will affect the overall completion time. For example, if you take half a minute to fill the kettle, the total time will go down to $7\frac{1}{2}$ minutes. On the other hand, if you reduce the time to get the cups to 2 minutes, the project will still take 8 minutes, because this activity is not on the critical path. Note also that if we bought a better kettle that could boil the water in 3 minutes, then the critical path would change and the total time would become 6.5 minutes.

Thus the critical path analysis allows us to determine the shortest time of completion of a task. It also helps to identify those critical activities that should be concentrated upon if overall completion time is to be reduced. So it is basically an optimisation technique.

Dynamic Programming

This is a method of solving multistage problems by means of sequential decisions. Problems with many variables can be greatly simplified by this

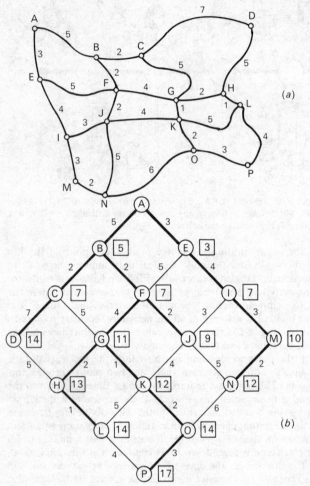

Fig. 6.26. Dynamic programming—determining the shortest path between points on a transport network. The shortest path from A to P is ABFGKOP, a total distance of 17 units.

technique, for if there are *n* variables then the problem can be reduced to *n* decision stages, each only dealing with one variable. Network analysis is one of the many engineering applications where dynamic programming is invaluable and we shall look at an example in this area.

In our earlier discussions of modelling we referred to the use of the digital computer for directing vehicles through a network of routes. Let us see how dynamic programming can help us to find the shortest route through such a network (Fig. 6.26(*a*)). Unlike Fig. 6.10, we shall assume that there is two-way traffic on all parts of the network.

Fig. 6.26(*b*) shows the decision tree for this example. There are 16 stations, and therefore there will be 16 decision stages. We shall look for the shortest route from A to P.

The following procedure is adopted:

1. Starting with A, we note it is connected to B and E. The minimum distance from A to B is 5 and this is recorded in the square beside B. The minimum distance to E is 3, and this too is recorded. Since route AB and AE are minimum paths they are marked by a heavy line.

2. We now wish to find the minimum distances from A to those stations connected to B and E. These are C, F and I. For node C there is only one sensible possibility, $5 + 2 = 7$, and this is recorded at C. (Note that it is possible to get to C by the route AEFBC, but this is not worth considering.) For node F there are two possibilities—routes ABF and AEF. Route ABF totals 7 and AEF totals 8; hence ABF at 7 is the minimum distance from A to F. This is recorded and the path BF is marked by a heavy line. For node E there is again only one sensible possibility, AEI = 7.

3. Having found the shortest routes from A to B, E, C, F and I, we now move on to look at those stations connected to C, F and I. These are D, G, J and M, and it is left to the reader to confirm the results shown on Fig. 6.26(*b*).

You will notice that by making sequential decisions we have reduced the alternatives to a maximum of two at each stage. This is the power of the method.

When the tree is completed we have the answer that the shortest route from A to P is ABFGKOP, with a total length of 17 units. But in addition we have the shortest routes from A to *any* other station. For example, to get from A to N, the route AEIMN should be used. You might argue that paths such as CG, DH, LP and so on should be eliminated since they are not included in any of the shortest routes from A, but remember that in a computer-controlled transport system we would also have to take account of journeys starting from *any* point on the map. So we would have to produce a similar decision tree for every station as starting point. For example, with C as the starting point of a journey, clearly the shortest route to G would be the path CG which was not part of any of the optimum routes from A.

6.12 Exercises

Answers to numerical questions are given in Appendix 1.

1. (*a*) What are models? (*b*) Describe the three major categories of models. (*c*) How do models assist the technologist?

2. (*a*) What dangers are there in using models? (*b*) What types of models are represented by paintings, maps, toys, statues, graphs, equations, definitions?

3. Construct as many models as you can for each of the following (five models is a good score): a resistor, a city transport system, sodium chloride, the solar system, a rocket, the atom.

4. A 5000 kg truck starts from rest on a horizontal surface. The engine supplies a thrust of 4000 N at the rear wheels, and the friction forces are 0.02 of the truck's

weight. Determine (*a*) the truck's acceleration, (*b*) the time it takes the speed to increase to 30 m/s and (*c*) the distance travelled during this time.

5. (*a*) What are analogues? (*b*) Draw two electrical analogues that would allow you to predict the acceleration of a rocket of mass M in response to a force P that increases linearly with time (i.e. $P = \mathrm{k}t$).

6. (*a*) Construct a model that illustrates the effects of births and deaths on population. (*b*) If the present world population growth rate is 2.1 per cent per year, how long will it take the population to double (see Chapter 10)? (*c*) In a town of population 500 000 the birth rate is 1.5 per cent per year and the death rate is 1.0 per cent per year. Assuming that these rates remain constant draw graphs for (i) the growth of the population over the next 100 years, assuming zero death rate, (ii) the decline of population assuming zero birth rate, and (iii) the variation of population taking both births and deaths into account. How long will it take the population to double?

7. (*a*) What are empirical models? Why are they necessary? (*b*) Describe three examples of empirical models. (*c*) An empirical model for traffic in a town is

$$T = 0.1P - 0.4H + 0.6E + 0.5C$$

where T is the daily number of trips taken from home to work and back, P is the population of the town, E is the number of employed residents, C is the number of cars owned and H is the number of households.
(i) Determine the present number of trips if $P = 500\,000$, $H = 150\,000$, $E = 200\,000$ and $C = 100\,000$. (ii) Determine T 20-years hence if the population growth rate is 2 per cent per annum and if it is anticipated that the employment rate will have dropped to 30 per cent of the total population. Assume that C and H maintain their present ratios to the total population.

8. (*a*) Give three examples of non-deterministic systems. (*b*) A man wishes to purchase a particular book and it is known that only 1 shop in 36 stocks this book. If the man visits 30 bookshops at random it can be calculated that his probability of finding the book is 0.57, i.e. in a thousand such searces of 30 shops he will probably find the book 570 times.

You can simulate these searches by using two dice, for the chance of throwing two sixes with two dice is 1 in 36; the same as the chance of finding the book in a given shop. A simulation of a single search requires the two dice to be thrown 30 times, successes being represented by the occurrence of double sixes.

Simulate 25 such searches and see how the number of successful searches divided by the total number compares to the theoretical value of 0.57. Increase the number of simulations to 50 to show that the result gets nearer to the theoretical as the number of simulations increases.

9. (*a*) A farmer wishes to enclose a rectangular area with an electric fence 400 m long. Representing the length of one side by x, draw a graph of area against x. Determine the maximum area and confirm this by differentiation. (*b*) A manufacturer makes cylindrical open-topped water reservoirs from plastic sheet. What is the maximum volume of water that can be held in such a reservoir if 120 m² of plastic are available.

10. A small factory manufactures record players (r) and televisions (t). There are three departments A, B and C with sufficient staff and machines to cope with a maximum of 12 000, 8400 and 14 400 hours of work per annum. During manufacture each record player requires 8 hours work in A, 4 in B and 12 in C. For the television the times are 10 hours in A, 12 in B and 16 in C. Each record player brings in a profit of £20 and each television £40. What mix of products should be manufactured each year for maximum profits? Are all departments worked to full capacity?

11. An engineering company designs a product in 20 weeks. Before the product can be assembled it is necessary to procure the six component parts. Part A has to be bought in, delivery taking 5 weeks. Parts B and C are made in the company workshops and require respectively 3 weeks and 2 weeks to complete. The remaining parts are already in stock. When all the parts are available it takes 2 weeks to

assemble them. Testing of the completed product takes a further week. After testing, the product is painted and packaged in 1 week, while simultaneously the paperwork necessary for exporting the product of Nigeria carries on, but takes 2 weeks.

Draw a network showing the various activities and determine the critical path from conception of the idea to despatch of the product. What operations should be speeded up in order to shorten the overall time?

12. Fig. 6.27 is a map of a region showing towns and roads; distances are in kilometres. (*a*) Use dynamic programming to find the shortest route from A to F. (*b*) If road JG is impassable, is the shortest route affected?

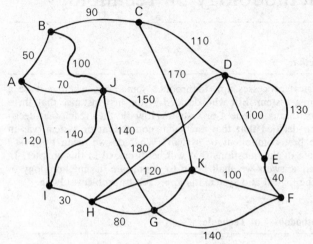

Fig. 6.27. A road map showing 11 towns and 20 interconnecting roads (Exercise 12).

6.13 Further Reading

Battersby, A., *Mathematics in Management*, Penguin Books, 1966.

Chow, W. C. W., *Cost Reduction in Product Design*, Van Nostrand Reinhold, 1978.

French, M. J., *Engineering Design: the Conceptual Stage*, Heinemann, 1971.

Harper, W. M., *Operational Research*, Macdonald and Evans, 1979.

Hingley, W., and Osborn, F., *Financial Management Made Simple*, Heinemann, 1983.

Meadows, D. H., Meadows, D. L., Randers, J., and Behrens, W. W., *The Limits to Growth*, Pan Books, 1972.

Meek, R. L., *Figuring our Society*, Fontana/Collins, 1972.

Wilson, W. E., *Concepts of Engineering System Design*, McGraw-Hill, 1965.

Woodson, T. T., *Introduction to Engineering Design*, McGraw-Hill, 1966.

7
The Methodology of Technology

7.1 Introduction

All technological processes aim to produce something, whether it be a component or a system. But who decided in the first instance that there was a need for that component or system? How did they reach that decision? And who decided that that particular component or system was in that case the best component or system? What do we mean by 'best'? These are some of the questions that will be discussed in this chapter. In providing their answers we shall begin to appreciate the methodology of technology, the logical and systematic approach to problem-solving.

7.2 The Methodology of Technology

We touched briefly on this topic in Chapter 1, but now we need a closer look. There are several steps in the procedure. First a **need** is identified; for example, there is a need to make cars more fuel-efficient. Secondly, **alternative means** of meeting this need are proposed and in our chosen example these could include (*a*) synthesising new fuels, (*b*) making cars lighter, (*c*) installing a rate of fuel consumption gauge on the dashboard, (*d*) improving engine design. The third step is the **decision-taking** stage, where the alternatives are compared and the best one chosen. Next the chosen solution is implemented and then tested.

This methodology is illustrated in Fig. 7.1, which shows that several feedback paths exist. For example, when a prototype of the chosen solution has been manufactured, initial tests may show it to be deficient in some respect and this may require the use of one of the earlier alternatives. We shall look at each stage of this methodology in some detail—first the perception of the need.

7.3 Basic Needs

From the earliest days Man's **basic physiological needs** have led to technological developments. He needed food; hence the development of agriculture and fishing. He needed a comfortable environment; hence clothing, housing, heating and lighting. He needed a means of defending himself

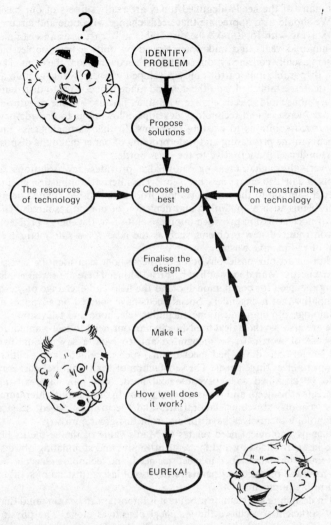

Fig. 7.1. The methodology of technology.

against wild animals and enemies; hence weapons. Another basic need was the perpetuation of the race, but this could be assured if food, shelter and defence were available.

Many products and systems available today are intended to meet these basic needs. Look at food, for example, and consider the array of gadgets aimed at assisting the housewife to prepare a meal. These include electric can openers, food mixers, desiccators, microwave ovens. Consider also the impact of technology on agriculture—hydroponics allows vegetables to be grown without soil and robot tractors are now being developed.

Thus many of the needs identified today are really subsets of our basic needs. We should also appreciate that needs change with time and circumstance. Staying with food, take the disposable milk carton as an example. A few hundred years ago milk was enjoyed by quite a few people, but purchasers usually collected it directly from the cow or from the farm. The need for disposable milk cartons only developed when the value of milk as a food became established, and the demand made it necessary to distribute it to every household. Needs change with time, because as time progresses knowledge increases and technology develops, allowing us to provide new and improved solutions to existing problems. In this particular case improvements in the processing and waterproofing of paper made the disposable carton a useful alternative to the glass bottle.

However, although increasing knowledge provides new solutions to problems, it can also create new problems. The internal combustion engine provided a solution to the transportation problem but it has created many other problems such as pollution and depletion of our oil resources. The science of nuclear fission provided another solution to the energy problem, but has introduced new problems such as the need for a safe method of disposal of radioactive waste.

In addition to our basic physiological needs, one can identify several secondary needs. Man does not live by bread alone! There are **social needs**, including the need for companionship and the help and exchange of views and sympathy that it engenders. Social needs have spurred on advances in the technology of communications, including telephone and television.

There are also **psychological needs**, including our need to feel wanted, to feel a sense of worth, to be recognised and to have a say in our own destiny. Society has developed mechanisms, such as religion and politics, to attempt to meet these needs. The satisfaction of Man's natural curiosity can also be classified as a psychological need. Man's problem-solving nature creates this need, and it has led to the growth of transport, literature, libraries, schools, museums, laboratories and the media. Indeed, science and technology themselves have sprung from this great curiosity.

Another psychological need relates to Man's value of the aesthetic; his desire to be surrounded by and to create attractive and stimulating objects. Art and music have developed to meet this need, and technology has played its part from the provision of paints to the use of lasers (instead of styli) in modern hi-fi equipment.

Thus in designing a system or product it is important to bear in mind that it is not sufficient to achieve the technical objectives alone. The physiological, social and psychological needs must not be forgotten.

Take, for example, a hi-fi system. Clearly there are functional requirements such as the quality of the recording and reproduction, and these can be specified in technical terms. One would find it difficult to identify a physiological need to be met by such a system, but there are many psychological needs to be taken into account. There is the feeling of self-esteem that ownership of such a system would nurture. Culture, taste and wealth could be reflected to the owner. There is the aesthetic pleasure given by the appearance of the system and by the music it generates. These factors have to be taken into account in the final design. A black box with high technical excellence would not be an adequate solution. Attention would have to be given to the style and colour of the system and, within a fixed cost, this may require some relaxation on the technical performance. This art of compromise between various needs is at the heart of the design process.

Needs and Demands

When is a need a need? Who says it is a need? Many so-called needs are the whims of individuals, and an all-embracing catalogue of such can be found in the records of the patent offices. Every week about 750 patents are registered for new inventions in the UK, and there are now over one and a half million of these inventions logged. In the US Patent Office, there are over four million patents. This indicates that lots of people have thought that many needs had to be met. Some of these needs were:

1. For a hat with a gun on top (BP 100,891).
2. For motion pictures with synchronised odour emission (BP 807,615).
3. For an apparatus for enabling the doors of poultry houses to be opened by the fowls within (BP 1700/1906).
4. For a device to put baby to sleep by means of periodic pats upon the rump or hind part of the baby (USP 3552.388).
5. For a sanitary garment for parakeets (USP 2882.858).
6. For a spirit level to allow sideburns to be correctly aligned (USP 2786.477).

None of these devices succeeded in attracting a market. And the market is the test. If sufficient people agree that there is a need for a product then its production becomes viable. In the business world it is generally agreed that if people desire a product enough to pay more for it than it costs to produce, then they should have it. But need and demand have to be distinguished. There may well be a demand for an electrically operated toothbrush, but is there really a need for such a gadget? It could be argued that much of today's advertising is aimed at creating a demand where there really is no need. This raises basic considerations of topics such as the quality of life, freedom of choice and so on, and these will be dealt with in more detail in Chapter 10.

Identifying the Real Need

Since the perception of a need is the starting point of the technological method, it is extremely important that a **real need**, and not an **apparent need**, has been identified. It is also important not to confuse the statement of the need with the possible means of satisfying the need. Consider the following examples:

(a) There is a need for a system of motorways to handle the traffic of the City of Belfast (population 450 000).
(b) There is a need for a bridge across the River Wet.
(c) There is a need for a wooden picnic hamper.

In order not to inhibit a breadth of outlook in the technologist's approach to solving these problems, they would be better stated as real needs:

(*a*) There is a need for a transportation system for the City of Belfast (present population 450 000; projected population in AD 2000—500 000).

(*b*) There is a need for a method of conveying people across the River Wet without getting them wet.

(*c*) There is a need for a portable container for food, crockery and cutlery, sufficient to provide a meal for four people.

The original statements of these needs assumed particular ways of meeting the real needs, i.e. (*a*) motorway, (*b*) bridge, (*c*) wooden container. These are not necessarily the best solutions to the problem. Several solutions are possible:

(*a*) Railway network; moving pavements; underground network; new buses, etc.

(*b*) Tunnel; boat; detour; helicopter, etc.

(*c*) Plastic; wickerwork; aluminium, etc.

Thus by recognising that the needs as originally stated were not the real needs, it is possible to propose many more solutions, thereby increasing the chances of finding an optimal solution.

7.4 The Search for Alternative Solutions

Once a need has been clearly identified, then the next step is to find a way of meeting it: the need has created a problem and it is necessary to seek a solution to the problem. In technology there are often many possible solutions, and it is necessary to decide which is the best. Clearly our chances of arriving at the best solution will increase as the number of proposed alternatives increase.

How can we increase the number of alternatives? How do we arrive at solutions? There are some basic rules that should be followed at this creative stage. First the designer or inventor should clear his mind of preconceived ideas concerning the solution, for this can severely restrict the number of solutions. In particular, existing solutions should not be overrated. Take the example of the science teacher who wished to test his pupils' knowledge of the atmosphere. He asked them to measure the height of a church tower using a barometer, assuming that they would note the change in atmospheric pressure at top and bottom, and hence deduce the height. He was surprised by the following solutions:

(*a*) Drop the barometer from the top and measure its time to fall; hence determine the height.

(*b*) Lower the barometer by string from the top; measure the length of string.

(*c*) Lower the barometer by string from the top, and when it is just about to touch the ground swing it like a pendulum; calculate the height from the period of oscillation.

(*d*) Compare the length of the barometer's shadow with that of the tower; knowing the length of the barometer, calculate the height of the tower.

(*e*) Persuade the verger to give the information in exchange for the barometer.

This example illustrates another important basic: the inventor must have a wide knowledge base. Solutions (*a*), (*c*) and (*d*) above required a knowledge of mathematics, physics and geometry.

Knowing Enough About It

Thring has argued that the problem-solver calls upon three knowledge bases: the intellectual brain, the emotional brain and the physical brain—or, as he aptly puts it, head, heart and hands. Technological problems usually require all three elements to be taken into account, and it is a criticism of our modern technological education that it concentrates on the head, to the detriment of heart and hands. Pupils and students are trained in analytical techniques, but know little about human relations and emotions, or craft-orientated skills. Looking back at our 'barometer' problem, one could argue that solutions (*a*), (*c*) and (*d*) were 'head' orientated, whilst (*b*) and (*e*) were respectively 'hand' and 'heart' orientated.

The knowledge base can be expanded by the inclusion of a variety of experts in the design team. In relation to the specific problem under investigation, it can be expanded by reference to the world's vast storehouse of experience. There are books, magazines, catalogues, records and patent files to be examined, and nowadays there are many agencies that provide computer-based information services. By this means you can at least be sure that you are not wasting time and effort in 'reinventing the wheel'. You may also find solutions that were proposed years ago but were not viable then because appropriate materials or production methods were not available. Perhaps they are now! But having gleaned all this information it is important to remember that it has been gathered to expand our alternatives, not to restrict them by providing preconceived notions.

The **world of nature** also provides a great deal of useful knowledge for the designer, and many great inventors have this source to thank for their inspiration. It is said that Marc Isambard Brunel conceived his scheme for constructing a tunnel under the Thames after observing that a shipworm constructed a tube for itself as it tunnelled through timber. The operation of the human ear gave Bell the idea of the telephone. The structure of the royal water-lily was the key to the construction of Sir Joseph Paxton's Crystal Palace. Observation of the humble wasp whilst nest-building inspired Réaumur to develop a new technique for paper manufacture. And there are many other examples, some true and some apocryphal like Newton and the apple, Watt and the kettle and Archimedes and his bath.

Man has still a lot to learn from nature. Our high-tensile steels are inferior to spiders' silk as far as strength/weight ratio is concerned. The ability of plants to perceive very small luminous intensities is astounding—the tips of the sweet pea seedling can detect the presence of a 100 watt lamp at a distance of 44 miles. There are many such examples of engineering ingenuity in nature, and the wise designer does not ignore this source of knowledge. But as an aside, you will note that Man has not always aped nature. Our aeroplanes do not use flapping wings and we rely a lot on the wheel, a device that nature seems not to favour.

Brainstorming

So the design group are now ready to have a knowledgeable and uninhibited attempt at solving the problem. How should they go about it?
There are several techniques to assist in the generation of alternatives. One
of these is brainstorming, where a group of people (5—15) sit together and
toss the problem around in an uninhibited fashion. This technique originated in the advertising field, where it was used to bring forward new ideas
for promotion of products. It has some basic rules:

(*a*) The problem has to be clearly defined.
(*b*) All ideas for solutions are welcomed, even if they seem ridiculous at
the time.
(*c*) As many ideas as possible should be collected.

For example, consider this problem. A company has found that its car
park is too small. What can be done? A freewheeling session could come
up with the following ideas:

1. Expand the car park.
2. Build a multistorey car park.
3. Build an underground car park.
4. Only allow small cars in.
5. Only allow senior staff in.
6. Reduce the workforce.
7. Operate a shift system.
8. Provide a bus to collect staff.
9. Provide taxis to collect staff.
10. Provide bicycles for staff.
11. Only employ people who live within walking distance.
12. Do not allow people to park if they live within walkable distance.
13. Pay for the use of public transport.
14. Charge a parking fee.
15. Store cars on top of each other.
16. Store cars on their ends.
    Etc.

Judgement on these proposals is deferred until the group has exhausted
its ideas. Only then are the alternatives criticised and refined or eliminated.
If, for example, the company did not wish to spend money, then only
solutions 4, 5, 6, 7, 11, 12 and 14 would be examined further. Of course
industrial relations are very important nowadays, and this would be a
strong influence in deciding which, if any, of this subset of solutions was
acceptable. None of them may be acceptable, in which case the constraint
of no expenditure would have to be lifted and replaced by the criterion of
minimum expenditure.

The brainstorming session has also redefined the problem. The basic
problem is how can people get to and from work at least expense and in
the most comfortable and convenient way.

Brainstorming is intended to free us from the shackles of conservatism.
A newspaper sponsored a contest for the best answer to the following
problem: Assume that in a balloon there are three famous men who have

made invaluable contributions to mankind. The first one made an important medical breakthrough, the second invented the microprocessor and the third is a renowned nuclear physicist. The balloon runs into a storm and can be saved only if one of the passengers is thrown overboard. Which man should be sacrificed? The paper received innumerable lengthy replies citing the merits of each man. But the judges awarded first prize to a 12-year-old whose answer was: 'The fattest one.'

Now that was a non-conservative solution, one that Alex Osborne would have liked. He was one of the earliest researchers to propose ways in which the mind could be freed from preconceived notions and accepted practice. He drew up a **check list** of questions for the would-be designer. The list, given below, is intended to stimulate the mind to examine different ways of expanding and rearranging ideas:

1. *Put to other uses?* In what way can an existing product or material be used differently—either as it is or after modification?

2. *Adapt?* What other product of Man or nature satisfies a similar problem? In what way can it be copied?

3. *Modify?* Can we change the shape, colour, motion, odour, sound and what effect does this have?

4. *Magnify?* What if it is made stronger, higher, faster, greater frequency, extra value, in greater quantities, exaggerated?

5. *Minify?* The reverse of magnify, even to the extent of eliminating?

6. *Substitute?* Substitution of people and things—skills, materials, processes, power sources, locations, methods of approach?

7. *Reverse?* Can the machine be run backwards, upside down, stop the moving, move the fixed?

8. *Rearrange?* Interchange components, other sequence, transpose cause and effect?

9. *Combine?* In what way can products, materials be combined? Extract the best features and combine these?

In attempting to answer these questions the designer is forced to look at the problem from many different aspects. It is hard work, but experience has shown the list to help the process. And the high degree of self-discipline and concentrated thought required cannot be overemphasised. Even the great Edison had to lock himself away in a tiny cupboard for many hours in order to develop and focus the intense mental effort required to invent the lamp filament.

This check list can be used when a new system or product is being created, or when an existing one is being developed to attract a greater market. Returning to our car-parking problem, we can see that most of the solutions produced by brainstorming could have been generated from the check list. For example, operation of a shift system comes under the heading of rearrangement—question 8. The proposed use of bicycles comes under the heading of adaptation—question 2. Reducing the workforce is an example of minification—question 5.

In attempting to get a bigger share of the market, a manufacturer of vacuum cleaners might conceive of a silent, cube-shaped, clear plastic, remotely controlled machine that perfumed the air as it operated and could also shampoo the carpet. Which of the above questions might have prompted the introduction of these different properties?

Morphological Analysis

The technique of morphological analysis provides another means of assisting the designer at the creativity stage. Like the check list it is to some extent a book-keeping exercise, to ensure that parameters and solutions are not overlooked, and that preconceived ideas do not dominate. Morphological analysis is often used as a follow-up to a brainstorming session. Imagine, for example, that a large airport authority had asked you to design a means of transporting their staff quickly and comfortably between the various areas of the airport complex. You organise a brainstorming session that generates a proposal for some form of mini-car, and you now wish to examine this possibility more closely. This is where a morphological analysis can help. In order to apply this technique, it is necessary to identify the important parameters involved, and then to list the alternative ways of achieving the desired results. In the case of the mini-car these parameters would be passenger position, vehicle support, power source, speed control and direction control. The various possible ways of meeting these requirements can be stated in matrix form below:

Passenger position	Vehicle support	Power source	Speed control	Direction control
Standing	Wheels	Human	Automatic	Manual
Seated	Castors	Electricity	Manual	Tracks
Prone	Air	Compressed Air		
	Magnetism	Petrol		

There are many possible overall solutions obtained by combining these particular solutions (in this case $3 \times 4 \times 4 \times 2 \times 2 = 192$). For example, a possible vehicle could have standing passengers and would ride on an air cushion, be driven by electric motor, have automatic speed control and be guided along a track. The method helps to focus attention on the parameters as well as on the solutions, and thereby helps the designer to recognise the problem more clearly. Do you think, for example, that stability should have been included as a parameter? And it is also worth noting that this technique is just another step in the design process. Further refinements are necessary—if we decide on an air cushion, what pressure will be required? If wheels are to be used, should they be rubber or steel? If human power is to be the driving force, will it be transmitted through pedals (bicycle) or by pushing with the foot (scooter)? If electricity is to be used will it be stored in a battery or collected from a wire or rail?

These techniques are aimed to assist the designer in his task, but technique is not enough; knowledge, effort and aptitude are required. A lot of work is needed to collect information and to develop experience and, as was mentioned earlier, a great amount of self-discipline is required from the designer. And even all of this does not produce a great designer or inventor—the flair for creativity must be there.

Creativity has been defined as a talent for discovering combinations of principles, materials or components, which are especially suitable as solutions to the problem in hand. Many people contend that it cannot be taught, but it can be assisted and nurtured, and this is the basic function of the techniques described earlier.

7.5 Elimination of Non-starters

The part of the design process that we have just described is basically an expansive process concerned with listing as many alternatives as possible. Now we have to go in the other direction. We need to converge in a discriminatory fashion towards the best solution.

Before studying closely how we arrive at the best solution, there is a basic step to be taken: the elimination of those alternatives that would not be feasible for one reason or another. In the terminology of the design process, we are looking for those alternatives that have exceeded certain **constraints**. These constraints can be of different types involving consideration of human, legal, economic, technical and time factors.

Human Constraints

All components and systems have to interface with humans in some way or other; after all they were conceived to meet some human need. Thus at the most basic level it is important to ensure that the interface between the human and the machine or system is acceptable to the human. Besides making a machine aesthetically pleasing (meeting a psychological need), we must ensure that it can be comfortably operated by a human. In fact a science, that of **ergonomics**, is devoted to this study. We now know, for example, how to lay out a control panel so that operator fatigue is minimised. Tables are available from which the designer can determine the dimensions and physical capabilities of the average human being, and this sort of information is very necessary when human operators are involved. Think of the vast number of devices that we interact with in this way— furniture, tools, televisions, typewriters, clocks, cars and so on. Would you buy a clock that required great force to wind up, or a car that had very heavy steering? On the other hand, would you buy an encyclopaedia whose print was too small or a smoothing iron whose handle got too hot? Or would you be happy to buy a lawnmower that was very noisy, or to work in a factory that had a very humid atmosphere? All of these examples illustrate that the human condition places limits on our design. Violation of any of these ergonomic, physiological or psychological constraints condemns a solution to the waste-paper basket.

Legal Factors

Many rules and regulations are necessary to ensure the smooth running of our society. There are, for example, **laws** regarding environmental pollution and safety standards, and indeed many of these embody some of the concerns we referred to above in our discussions of human factors. These are inviolable. So forget about that internal airport transport system that uses petrol driven vehicles if the exhaust fumes are going to poison the air!

The technologist must also be careful not to infringe **standards and codes** where they exist, and where local authorities or government have asked for their application. These can operate at several levels—company, industry, local, national and international. At the top, the International Standards Organisation (ISO) has produced, amongst others, standards for drawing paper dimensions, for screw threads, for engineering units. And there is a lot to be said for the use of standards, for their use facilitates inter-changeability of components and ensures uniformity of properties, and ready access to books of standards relieves the designer of some of the more mundane aspects of the design process.

Whilst standards are concerned with dimensions of components and composition of materials, codes are more directed towards the definition of good practice. They cover recommended design procedures for items such as gears and welded joints, and recommend application procedures such as the preservation of timber and the use of concrete.

On the legal front, the designer should also be aware of the **patent laws**, and if any of the proposed alternatives infringes an existing patent it should be viewed with great suspicion. If this constraint is not treated with due respect at an early stage, a company could spend a lot of money on design, development and production, only to find itself on the receiving end of an action from a patentee who had thought of it all years ago.

Finally the whole business of **contracts** and agreements presents the designer with another type of legal constraint. A contract obliges both parties to agree to satisfy a set of defined conditions. For example, A will give B a sum of N pounds if B provides X items with performance Y within time Z. Once the contract is signed the designer has to interpret X, Y and Z as constraints and any solution that does not have potential for producing a minimum of X items, each with a minimum performance Y, in a maximum time Z, should be discarded or carefully reviewed.

Economic Factors

Given enough money and time there are few things that Man cannot accomplish. But where does the money come from? For the majority of items, we, the consumers, have to supply the money when the item is bought over the counter. And there are many large-scale capital projects, such as hydroelectric schemes and motorways, where we, the tax payers, have to foot the bill. In the case of the consumer product the amount of money available for the project is, in the end, largely determined by what the consumer is prepared to pay for the item. Monies available for the big projects are affected by political attitudes. But whatever project the designer is involved in there will be a financial constraint to be considered.

Whatever alternative the designer is considering, he should be sure that there is enough capital available to finance the project right from the start of the design until the component or system is sold and the investment recovered. He should not fail to recognise that the pound, dollar or franc is just as significant an engineering unit as the kilogram or the metre.

The total cost of a project includes many parts. There are overheads to be considered including such things as taxes, insurance, heat, light and power. There are production costs including design, manufacture, packaging and maintenance. And there can be a considerable expense involved in transportation of the product to the consumer. We shall hear more of this in Chapter 9.

Occasions will arise when an alternative solution blatantly violates a financial constraint. For example, in order to get 20 people a day across a river, would you advocate a rowing boat or a tunnel? Other proposed solutions may require expertise or production facilities that the particular manufacturer does not possess, and calculation may prove that the monies required to make up these deficiencies are excessive. But it has to be emphasised that in most cases it is no simple matter to make a reliable prediction of economic feasibility. Nevertheless, it is an essential part of the designer's work, and we shall refer to it again later in this chapter.

Technical Factors

Here we have to be sure at the outset that none of the laws of physics is being violated. We do not wish to repeat the mistakes of the advocates of perpetual motion who paid no attention to the first law of thermodynamics. We do not wish to be involved with proposals for lifting one's self by one's boot straps. Nor can we retain solutions that require non-existing technologies. For example, although it may be desirable to send people across the River Wet by disintegration and reassembly of their molecular structure, it is clearly not on at present.

There are many other less sophisticated technical constraints. For example, if our proposed new cubical vacuum cleaner turns out to be too big to pass through the doorway, then we shall have to reject or modify the solution. Other obvious technical constraints are determined by **standards** and by **regulations**. For example, electrical equipment designed for the home usually needs to be compatible with the domestic electricity supply. We also have to be sure that our product does not violate the pollution or safety laws that we referred to earlier. It is often said that it is not sufficient to make a product that is foolproof nowadays—it has to be idiot-proof!

When assessing the technical feasibility of a proposal, the designer will usually make heavy use of the modelling techniques referred to in Chapter 6. Pictorial, physical, analogue and mathematical models are essential for determining whether a particular solution will meet a technical specification.

7.6 Choosing the Best Solution

Having identified and rejected those solutions that violate known constraints, we now have a short-list of viable alternatives from which we must choose the best. What is the best solution? The 'best' depends on the individual and the circumstances. The user of a new product will want to know what **benefits** he can derive from its use, and how much these benefits will **cost**. In some cases it may be required to achieve the greatest benefits for a given cost—I have £5 to spend on a bottle of wine: which should I choose? Or, at a higher level, a government has a million pounds to spend on hydroelectric power: which type of power station should it choose? In either case the solution giving the greatest benefits for the fixed cost would be considered to be the best solution. There will also be many cases where decisions are based on the least cost, and this obviously has to be done when money is scarce and a minimal benefit at least is essential. People on low incomes are often forced to buy the cheapest commodity, and in their case the cheapest is the best. However, in most cases people will be looking for the best buy, and will be seeking, usually unknowingly, to maximise the **benefit/cost ratio**.

On the other hand, the manufacturer or the technologist wishes to have a maximum return on his investment, and the product that achieves this will be considered to be the best. This basically is quite similar to the user's aim of maximum benefit/cost ratio, for a product that gives maximum benefit to a user is likely to sell well and provide a good return to the manufacturer. Similarly, the price that the user pays (his costs) for a product will be related to the investment costs accrued by the manufacturer.

So both user and manufacturer will want to maximise benefits and minimise costs, and usually the primary criterion determining which is the best solution is the benefit/cost ratio. The secondary criteria are the benefits and the costs. The manufacturer's benefits are related to the financial return. The consumer's benefits are dependent upon the attributes of the system or component. When you buy a car, for example, you look for the **attributes** of comfort, reliability, speed, appearance and so on. Costs, for the user, are reflected in the price tag, but for the manufacturer they include the costs of design, manufacture, distribution and maintenance. But, as mentioned above, the selling price and the manufacturer's costs are related.

Assessing the Benefits

Let us look at benefits more closely. How can we tell if one solution gives us a greater benefit than another? In order to help, imagine that you want to buy a briefcase and you have several manufacturers' catalogues and price lists in front of you. Where to start? What do you want from a briefcase? What benefits will you gain from owning a briefcase? Well, for a start I could list several attributes that I would expect a briefcase to have. It should have a good appearance; it should be well made and robust; the internal layout should be able to store a variety of articles, from pens to papers; it should be lightweight.

Now, how would you choose between a good-looking, poorly made case and a well-made unattractive case? In order to do this it is helpful to establish a **value scale**. How much do we value each of these attributes? Are some more valuable to us than others, and if so, by how much?

Let us use the following shorthand for the attributes in our example:

$$A1 = appearance$$
$$A2 = robustness$$
$$A3 = layout$$
$$A4 = lightness$$

In order to determine a 'batting order' of value it is useful to compare the attributes two at a time according to the technique described by Svensson. This is easily done by using the matrix below:

	A1	A2	A3	A4
A1	0	0	0	0
A2	1	0	0	0
A3	1	1	0	$\frac{1}{2}$
A4	1	1	$\frac{1}{2}$	0
Total	3	2	$\frac{1}{2}$	$\frac{1}{2}$

Moving down through column A1 (appearance) we firstly compare A1 (appearance) with A2 (robustness). My conclusion is that greater value should be placed on appearance; so a 1 goes in the table at that point. If the attribute in a column is less value than that in a row, then we enter an 0; for example, I decided that lightness was less important than robustness. If there is any indecision a $\frac{1}{2}$ should be entered. When the table is completed the scores in each column can be aded giving the **ranking order** of value V. In this case appearance comes first, robustness second and layout and lightness tie for third place.

It is now necessary to quantify this value scale, and one way of doing this is to allocate points as follows:

10 for the most important
8 for the very important
5 for the moderately important
3 for the slightly important
1 for the least important.

One can also interpolate within this scale. In the chosen example we could allocate points as follows:

$V1 = 10$ for A1: appearance
$V2 = 8$ for A2: robustness
$V3 = 5$ for A3: layout
$V4 = 5$ for A4: lightness

It must be emphasised that this ranking order and the values assigned were determined by me. It is very likely that you would have come up with a different listing. This is a very important point, especially when one is trying to devise some product to meet a need. Will the consumer's value scale be the same as yours? Most likely not, and it would normally be necessary to do a market survey to get some measure of the consumer's values.

Forgetting briefcases for the moment, let us compare the values that a designer might place on similar attributes in different artifacts. Take a vacuum cleaner component and an aircraft control system component. The designer's table of values could look like the following:

	Vacuum Cleaner	*Aircraft*
Lightness	3	8
Reliability	7	10
Appearance	8	1
Maintainability	5	5

This emphasises that all parts of an aircraft must have the highest reliability. And since an aircraft is built to carry a payload and not just itself, it is essential to keep the weight of its constituent parts to a minimum. The vacuum cleaner, like our briefcase, has to please the eye as well as clean the room, and so appearance is rated highly in its case.

We have now identified the important attributes of the artifact and have assigned values to each of them. The next step is to see how each of our alternative solutions performs with respect to these attributes. In order to do this we again have to attempt to quantify the problem by assigning numbers P (performance) to reflect the degree to which a given attribute is satisfied. The following table is useful:

$$P = 10 \quad \text{for complete satisfaction}$$
$$P = 9 \quad \text{for extensive satisfaction}$$
$$P = 7\tfrac{1}{2} \text{ for considerable satisfaction}$$
$$P = 5 \quad \text{for moderate satisfaction}$$
$$P = 2\tfrac{1}{2} \text{ for minor satisfaction}$$
$$P = 1 \quad \text{for minimal satisfaction}$$
$$P = 0 \quad \text{for no satisfaction}$$

Occasions will arise when a precise value can be given to P and the above table is not required. Assume, for example, you were looking for a car with a top speed of 180 km/h and the one on offer could only manage 150 km/h. In this case $P = (150/180) \times 10 = 8.33$.

Staying with our briefcase, let us assume that we are faced with three alternatives, and using the above table we are able to list the performance P with respect to the attributes A1 to A4 as follows:

Attributes	Alternatives		
	1	2	3
A1	10	$7\tfrac{1}{2}$	5
A2	5	10	$7\tfrac{1}{2}$
A3	5	$7\tfrac{1}{2}$	$7\tfrac{1}{2}$
A4	10	$7\tfrac{1}{2}$	10

For example, from an examination of the third alternative we have concluded that it is considerably robust, moderately attractive, its weight is just right and its layout gives considerable satisfaction. Thus, in addition to knowing how much we value each attribute, we now know how well each solution performs with respect to that attribute.

We now have enough information to determine which solution offers the greatest benefits. It is common practice to define the **benefit** B as the product of the value V and the performance P, i.e. $B = VP$, and the reader will best appreciate the significance of this definition of benefit by considering two extreme cases. First, if an alternative has a zero performance for a particular attribute, then no matter how highly we value that attribute there can be no benefit to the user. Secondly, no matter how high the performance, if we place no value on that attribute we will receive no benefit. The benefits of each alternative can be set out in table form:

Attributes		Alternatives		
		1	2	3
A1	Appearance	100	75	50
A2	Robustness	40	80	60
A3	Layout	25	37½	37½
A4	Lightness	50	37½	50
	Total benefit	215	230	197½

Thus we conclude that the second alternative gives the maximum benefit. It may not be as pretty as the first and it may be heavier than the first, but it will be well-laid out and will last a lifetime. So if we want the maximum benefit, the second briefcase is the best.

Counting the Costs

But benefits are only one side of the story. We must also count the costs! Let us assume that the first briefcase costs £50, the second £50 and the third £40. Clearly then, if the pursestrings are tight the third alternative, with least cost, is the best.

Thus, based on maximum benefit the second case is best, whilst based on least cost the third is the winner. We mentioned earlier, however, that most people would define the best buy as that which maximises the benefit/cost ratio.

Alternative 1	Benefit/cost = 215/50	= 4.30
Alternative 2	,, = 230/50	= 4.60
Alternative 3	,, = 197½/40	= 4.94

Thus using this criterion, alternative 3 would be the best buy.

Normally cost benefit analysis attempts to express both costs and benefit in money terms. In our example the costs are already expressed in pounds. For the benefits it is often assumed that the monetary value is directly proportional to the perceived benefit, i.e. $£B = kB$. But how do we determine the constant k? Well, since cost benefit analysis is concerned with comparisons we can assume that one of our alternatives has equal costs and benefits and then see how the others fare in comparison. For example, if we assume that costs and benefits are equal in monetary terms for alternative 2, then we have £50 = $230k$, giving $k = 0.217$. Hence we have:

	Benefit	Cost	Pay off
Alternative 1	215 × 0.217 = £46.66	£50	−£3.34
Alternative 2	£50	£50	0
Alternative 3	197.5 × 0.217 = £42.86	£40	£2.86

This confirms our earlier calculation that showed alternative 3 to have the best benefit/cost ratio.

It is interesting to note that one of the briefcase's four attributes, its appearance, is an intangible. If we wished we could assess the other three by scientific tests. Weight can easily be determined; robustness could be evaluated by a test rig designed to subject the case to various loads; layout could be specified in terms of capacity and ability to hold various sizes of paper, etc. But judgement of appearance is very subjective and cannot be measured on some scale or other. Nevertheless, the above procedure enabled us to put an actual money value on appearance; for the first briefcase the appearance turned out to be worth £21.70 (100 × 0.217) to me.

Cost benefit analysis is praised by many for removing subjectivism from decision-making and replacing it by scientific rationality. On the other hand, there are people who accuse it of attempting to measure the immeasurable, and their differing points of view often meet head-on at public inquiries when large public sector projects like motorways and power stations are proposed. If a new power station is being planned in the vicinity of a picturesque spot, how much would we cost the decrease in the beauty of the countryside? Would the benefits exceed this cost?

Costing such intangibles is difficult and controversial, but the courts have to put a money value on human life, and we now can put money values on travel time saved by building new roads and on the benefits of higher education. So at best, cost benefit analysis is a powerful method that eliminates subjectivism, and at worst it is a means of focusing on the relative pros and cons of a project.

The Consumer's and the Technologist's View of Costs

Now the above example concerned the choice made by a consumer; a choice from alternatives offered by producers. The technologist, on the other hand, is a producer and he has to choose among the various alternatives that he himself has generated. To the consumer the cost involved is represented by the money he has to pay, and the benefit by the satisfaction he receives from the product. For the technologist manufacturing the product, the cost is represented by the labour and capital he has to use for production, and the benefit by the income he gets from sales. As we argued earlier, the consumer's costs will be related to the manufacturer's costs, and the article which costs a lot to produce is likely to be expensive to purchase. Similarly, the greater the benefit to the consumer by way of satisfaction, then the greater the benefit is likely to be to the technologist in terms of sales. Hence both consumer and producer will wish to maximise the satisfaction to the consumer. Thus, if a manufacturer of briefcases conceived of three designs, the alternatives mentioned earlier, he would be seeking, by means of market research, to determine how much satisfaction the public would receive from each. Using this information, and knowing the costs involved in manufacture, the manufacturer could then do his own cost benefit analysis, and from this decide which briefcase to manufacture.

Now assessing the cost is easy for the consumer; it is on the price tag. The manufacturer has a more difficult task, and some of the factors he has to take into consideration have already been referred to. There are several methods by which he can compare the costs of products, but we will have to restrict ourselves to considering only two of them. They both assume that there is a reasonably clearly defined cash flow for the project, but we should remember that there are many cases where there is a considerable degree of uncertainty about future markets. If you are interested, you should refer to a good text on financial management.

Let us first examine a technique that evaluates the financial returns on the project but does not take into account its timing. Staying with our example, we will assume that a manufacturer knows the costs and returns or benefits involved in producing our three briefcases. He wishes to decide on the most attractive proposition as far as his company is concerned. In order to manufacture any of the briefcases it will be necessary to invest in special-purpose machinery; so there is an initial capital outlay. There will be other costs, such as labour and materials. The table below lists the capital cost and the net income per annum predicted for each alternative.

Briefcase	Capital cost (£)	Net income (benefit) (£)		
		Year 1	Year 2	Year 3
1	50 000	40 000	20 000	—
2	60 000	25 000	25 000	25 000
3	50 000	10 000	20 000	30 000

The **payback period** is often used as a method of investment appraisal. It is defined as the number of years required for the stream of cash proceeds generated by an investment to equal the original cost of that investment—in other words, the time required for the total benefit to equal the initial costs. Going back to our table and assuming that the rate of cash inflow is even throughout a year, then the payback periods, PB, are:

Briefcase 1: PB1 = 1.5 years
Briefcase 2: PB2 = 2.4 years
Briefcase 3: PB3 = 2.67 years

Thus this particular method of investment appraisal puts the first briefcase proposal at the top of the list.

The payback period is easy to compute and to understand and it emphasises cash flow as a dominant factor. It can, however, be criticised for ignoring income that arises after the payback period, and it is biased towards projects having higher cash flows in the early years. And it presupposes that money received at some future date is worth as much as money received now.

Let us now look at a technique that takes into account the effect of time on the value of money. We all know if we lodge a sum of money P in a bank at an interest rate r, the total A accumulated after n years will be

$$A = P(1 + r)^n \qquad (7.1)$$

So if we invest £100 at per 10 per cent, it will amount to £121 after two years. Now we turn this on its head and ask: if a project promises an income of P in n years, what is P worth to us now? For example, if we can predict that a product will bring in £121 after two years we can argue that its present value P is £100, for if we were to invest our £100 in a bank now, it would have grown to £121 after two years. This procedure of finding the **present value** of future cash flow is called **discounted cash flow**. In general, the present value P is given by

$$P = A/(1 + r)^n$$

where A is the amount due to be received or paid in n years from now, at a given discount rate r. By converting all future benefits and costs to present values, we can determine the present worth of a project, i.e. the present value of the benefit minus the present value of the cost.

This technique will become clearer if we apply it to our briefcase project. Assuming a discount rate of 10 per cent we have

Year	Case 1		Case 2		Case 3	
	Income	Present value	Income	Present value	Income	Present value
1	40 000	36 364	25 000	22 728	10 000	9 091
2	20 000	16 528	25 000	20 660	20 000	16 528
3	—	—	25 000	18 783	30 000	22 539
Gross present value		52 892		62 171		48 158
Cost		50 000		60 000		50 000
Present worth		2 892		2 171		− 1 842

For example, for briefcase 3, we predict a net income of £30 000 at the end of the third year and the present value of this sum is

$$30\ 000/(1.1)^3 = £22\ 539$$

The table shows that the first briefcase is the most promising venture. Briefcase 3 should be rejected because its present worth is negative; if the manufacturer had the £50 000 in capital he would get a better return by lodging it in a bank than by proceeding to manufacture briefcase 3.

Similarly, if he were to borrow the £50 000 over the three years, one can show that the income would not be sufficient to repay the loan plus interest (£66 550). You may of course ask why the manufacturer does not increase the price of the third case, thereby increasing his income. This, however, assumes that the same number of people will buy at the increased price, which is doubtful. We assume here that the marketing teams have investigated this fully, and that the prices have been chosen to give maximum return.

Thus, for this particular example, the first alternative comes out top, based both on payback period and on present worth. But this will not always be the case, and many occasions will arise where different techniques of investment appraisal will advocate different projects. This should not lead the reader to brand all the techniques as useless, for no one technique is the best for all purposes. They are useful, if for no other reason, in that the technologist will come to a clearer appreciation of the important financial features of the project whilst attempting to understand the differing recommendations of the various techniques. But we cannot do justice to the full power, complexity and advantages of financial management here.

So let us recap the saga of the briefcase. Given three alternative briefcases, I, the consumer, assessed their benefits and concluded that number 2 gave the greatest benefit and number 3 cost the least *and* had the best benefit/cost ratio. The manufacturer, on the other hand, faced with the decision as to which of the three cases to manufacture had also to look at cost and benefit. His benefit would be the income generated from the sales of the briefcases. He had to assess how the average consumer would value the particular attributes of each case, and then his marketing team had to predict how many cases would be bought, and when, at a price calculated to give the best return. The result of this study showed that the manufacturer would be wise to proceed with the manufacture of the first brief case.

7.7 Implementation of the Solution

The culmination of the design process is the implementation of the chosen solution, and before a product can be manufactured or a system put into practice, the technologist has to persuade his colleagues that it is worth doing. Although he may be convinced that there is a best solution he now has to convince others, and this is no easy task. Many engineers start out on their careers assuming that if their proposals are technically and economically superior they will be automatically adopted. They forget that they have to sell their solution to colleagues, clients and financial backers who may be wary about change and about risks, and this requires skill in the presentation of a case in written form, graphical form or by word of mouth. Sydney Love has pointed out that you cannot 'make people understand' your communication, but you can 'make it understandable'. The responsibility for successfully communicating a design idea rests with the sender.

A good piece of advice for most communication is to keep it simple. Where possible avoid technical jargon and use plain English. And when talking to management remember that they are basically interested in what it will do, what it will cost, when the money will be needed, and what will be the likely returns.

When the designer's arguments have been accepted, the next step is the physical creation of his design. He should remain in contact throughout this process, monitoring and detecting and remedying incorrect details. All being well, only minor modifications will be necessary, but occasionally when a prototype has been completed it may be found to have a serious shortcoming that requires another solution to be adopted. This feedback path is shown in Fig. 7.1.

Let's conclude this chapter on a humorous note. Two golfers, Harry and George are on the first tee:

Harry: 'I've just got one of these new fangled golf balls.'
George: 'Oh! What's special about it?'
Harry: 'Well, if you hit it off line a fin comes out and corrects its course. And if you do manage to land it in the rough it emits a bleep that helps you to find it. And on top of all that it floats on water too.'
George: 'Sounds great! Where did you get it?'
Harry: 'I found it.'

Clearly this was a design that did not live up to expectation.

7.8　Exercises

Numerical answers are given in Appendix 1.

1. Technology is concerned with problem-solving. Pick three problems that you are likely to meet throughout the day and show how the various steps of the methodology of technology are applied to their solution.

2. (*a*) Distinguish between needs and demands. (*b*) What are our basic needs? (*c*) What needs are satisfied by chocolate, television, shoes, digital watches, aeroplanes, butter, books, aspirin, chairs, bricks?

3. (*a*) Distinguish between real needs and apparent needs. (*b*) Rewrite the following suggestions in terms of real needs: (i) plastic garbage containers; (ii) money for the world's starving people; (iii) a high-speed train; (iv) margarine that tastes like butter; (v) a rust-proofing agent for cars; (vi) a better way of handling road maps in cars.

4. (*a*) What is creativity? (*b*) Serendipity is the faculty of finding interesting or valuable things by chance. List six inventions that were discovered this way. (*c*) List six inventions that owe their origin to observation of the world of nature.

5. Bring together a group of at least five people for a brainstorming session. Try to produce at least 20 solutions to each of the following problems: (i) helping a blind person across a busy road; (ii) reducing traffic jams; (iii) finding new uses for bricks; (iv) reducing the world's use of paper; (v) making cars safer.

6. A holder for a flagpole consists of a metal tube 1 m long, 50 mm inside diameter, sunk to a depth of 100 mm in concrete. A ping-pong ball, diameter 30 mm, falls down the inside of the tube. Suggest ten ways of retrieving the ball. Which is the best? And why?

7. List four important parameters of a lawn mower. Using morphological analysis, suggest a novel form of lawn mower.

8. A woman wishes to purchase a small powered bicycle (moped) to get her to and from work. She looks for four attributes in a moped: A1—she requires a top speed of 100 km/hr; A2—she requires at least 40 km from each litre of petrol; A3—she wants it to look attractive; A4—she wants it to be lightweight. She rates A2 most important, A3 and A4 very important, and A1 moderately important.

On visiting the local motorcyle retailer she finds three mopeds that could be acceptable: M1, M2 and M3. From the manufacturers' literature she is able to determine that the top speeds are:

M1	100 km/h
M2	80 km/h
M3	70 km/h

The fuel consumptions are given as:

M1	30 km/l
M2	34 km/l
M3	40 km/l

As far as appearance is concerned she finds M1 moderately satisfying, and M2 and M3 considerably satisfying. In terms of lightness she finds M1 considerable satisfying, M2 extremely satisfying and M3 completely satisfying.

The price tags on the mopeds are:

M1	£200
M2	£170
M3	£185

(*a*) Calculate the benefits the woman will receive from each moped. (*b*) What is the benefit/cost ratio for each? (*c*) Which moped should she purchase?

9. What do you believe were the important criteria that the designers of the following had to consider? (i) An interchange between two intersecting dual highways; (ii) an artificial hand; (iii) a factory for making televisions; (iv) a supersonic aircraft.

Identify criteria that appear to be conflicting and require compromises to be made.

10. A company plans to manufacture a device for automatic alignment of television aerials. Their designers propose three models A, B and C. Initial costs are estimated to be: A = £50 000, B = £60 000 and C = £80 000. The Marketing Department examine the potential market and provide management with the following projections:

Net Income (£)

Model	Year 1	Year 2	Year 3	Year 4
A	30 000	20 000	10 000	—
B	30 000	20 000	20 000	15 000
C	10 000	20 000	40 000	50 000

(*a*) Determine the payback period for each model. (*b*) Using discounted cash flow, find the present worth of each model. (*c*) Based on this information which model should management put into production?

11. Design an arrangement that will act as an alarm clock for someone who is totally deaf. Your design should be practical, reasonably economic, efficient and safe.

12. Discuss, with reference to the methodology, the design of a microcomputer. Refer to the knowledge base and indicate what branches of science are important. Identify the various constraints that apply.

7.9 Further Reading

Those marked with an asterisk are particularly recommended.

*Adams, J. L., *Conceptual Blockbusting*, W. W. Norton, 1979.

Hingley, W., and Osborn, F., *Financial Management Made Simple*, Heinemann, 1983.

Laithwaite, E., *Engineer through the Looking Glass*, BBC Publications, 1980.

Love, S. F., *Planning and Creating Successful Engineering Designs*, Van Nostrand Reinhold, 1980.

Meek, R. L., *Figuring out Society*, Fontana/Collins, 1972.

Meyer, J. S., *Great Accidents in Science That Changed The World*, Arco Publishing Co., 1967.

Osborne, A. F., *Applied Imagination*, Charles Scribner's Sons, 1963.

Pitts, G., *Technique in Engineering Design*, Butterworth, 1973.

*Svensson, N. L., *Introduction to Engineering Design*, Pitman, 1974.

*Thring, M. W., and Laithwaite, E. R., *How to Invent*, Macmillan Press, 1977.

8
Technology in Action

8.1 Introduction

Technology influences all of our lives. It is all-pervasive, and readers are challenged to look around to see if they can spot anything that is not a product of technology. I am lucky to live in the beautiful countryside of County Down in Northern Ireland, and the view from my study window is delightful. But it also presents a panorama of the technologies for I can see power lines (electrical engineering), telephone wires (communications engineering), houses (building, architecture, surveying), roads (civil engineering) and, cars (mechanical engineering). Even the fields and the cows are affected for modern agricultural engineering relies on chemical technology for fertilisers and on computers and control systems for optimising the milk yield of cows.

So the task of illustrating technology in action is a daunting one. Indeed, several encyclopaedias have been devoted to this topic and the reader is advised to refer to them. But for the purposes of this book I have chosen a different approach. Instead of attempting to compete with the encyclopaedias, this chapter will consider two specific areas of technology—**urban transportation** and **robots**. To some extent we are all familiar with the former, and this should help us to follow the developments in that area. Robots, on the other hand, are not quite so familiar to the average reader, and on the whole there is a considerable degree of misrepresentation and misunderstanding of this important area of technology.

8.2 Transportation

Transport has developed from the sled and bullock cart to high-speed trains, Concorde and the space shuttle. These latter can get us from city to city and from planet to planet with greater speed and comfort, but we still need improved methods of transporting people and goods within our cities. This is the urban transportation problem that faces many cities and now challenges the engineers.

The automobile

The automobile receives a lot of criticism today and there is no doubt that it has made its contribution to the urban transport problem. Consider its disadvantages:

1. Its engine is inefficient, wasting fuel and fouling the air.
2. The car wastes space on the roads, because it is rarely fully occupied.
3. The car needs to be stored at the end of each trip. In cities this means that streets are congested with parked cars, and that expensive urban land is used for multistorey garages and parking lots.
4. The car is a potential safety hazard and deaths on the road are increasing every year.
5. The congestion which it causes impedes the flow of public transport such as buses, and thus those people who cannot afford a car or who are too old or too young to drive, have to suffer.

But in spite of this depressing list of shortcomings, the motorcar becomes more popular each year. Hence in the eyes of the average traveller the car's advantages must outweigh this list of disadvantages.

What are its advantages?

1. It offers a door-to-door transport facility.
2. It is available for use at all times.
3. It is not restricted to particular routes, but can go almost anywhere at the whim of the driver.
4. At a more basic level it can give pride of ownership, and it allows one to travel anywhere whilst remaining in a completely private world.

These advantages have made the car the success it is today, and indeed have led to the very transport problems that technologists are attempting to solve. But should not the success of the motorcar be a lesson to transport planners? Is it not obvious that any future conceived means of transportation should be designed to incorporate as many of the car's advantages as possible? Let us see what could be done to reduce the car's disadvantages whilst maintaining its advantages.

The Town Car

It has been proposed that a reduction in the size of cars would be a step in the right direction. This should reduce the space taken both on the roads and in the parking lots. What savings could we expect from such a measure? First of all its is necessary to decide just how small we can make a car without reducing it to a toy.

Stability requirements must be considered when deciding on the dimensions and weight distribution of a car. For example, when a car brakes there is a tendency for the rear to lift; indeed, excessive brake forces and bad weight distribution can make it tip over completely. It is easy to see why this happens if we assume (Fig. 8.1(*b*)) that the rear wheels are momentarily off the ground. Three forces act on the car: (i) the road reaction N pushing upwards on the car; (ii) the car's weight W acting downwards; (iii) the brake force F (wheel friction) acting backwards. With respect to the centre of gravity G, the road reaction produces a clockwise moment Na tending to set the rear wheels back on the road again. On the other hand, the brake force produces an anticlockwise moment Fh

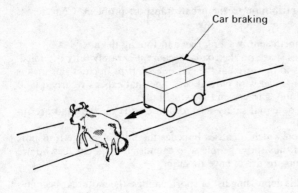

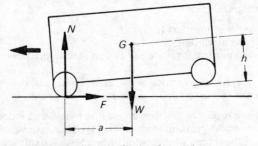

Fig. 8.1. During braking the friction force F between tyres and road tends to tip a car over.

which tries to tip the car over. So for stability it is necessary to ensure that the road reaction moment wins. Hence

$$Na > Fh \text{ for stability} \tag{8.1}$$

Now N can be assumed to be equal to the car's weight*, so equation (8.1) becomes

$$Wa > Fh \text{ for stability} \tag{8.2}$$

Furthermore, elementary mechanics tells us that the friction force can be assumed to be proportional to the reaction N, and hence to the weight W. Therefore

$$F = kW \tag{8.3}$$

where k is a constant called the coefficient of friction.

Equation (8.2) then becomes

$$Wa > kWh \tag{8.4}$$

or $$a > kh \text{ for stability} \tag{8.5}$$

This shows that the centre of gravity must not be too far forward (a too small) or too high (h too large).

*This assumes that at that instant there is no vertical acceleration of G.

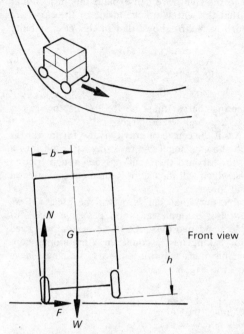

Fig. 8.2. During cornering, the friction force *F* between tyres and road tends to tip the car out of the turn.

So stability during braking places certain restrictions on the length and the height of the car. Stability during cornering, on the other hand, can be shown to place restrictions on width. Fig. 8.2 shows a car turning to the left, and the important thing to appreciate is that if the car is to move to the left there must be a force pushing it in that direction. This force *F* is provided by friction between the wheels and the road.

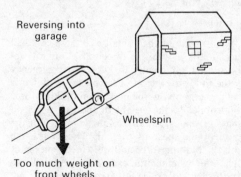

Fig. 8.3. Wheelspin can occur when reversing, if the centre of gravity is too far forward.

Now, it is common experience that during a turn to the left there is tendency for the car to tip to the right. We can explain this in a similar way to the above. Assume that the wheels on the inside of the curve are momentarily off the ground. Fig. 8.2(*b*) shows that in this case stability requires that

$$Nb > Fh \tag{8.6}$$

or
$$Wb > kWh$$

or
$$b > kh \tag{8.7}$$

Thus for stability in cornering the car must not be too narrow (*b* too small) or the centre of gravity too high (*h* too large).

Another hazard which arises if the centre of gravity is too far forward is illustrated in Fig. 8.3, which shows a small car reversing up a hill into a garage. If the weight is too far forward there will not be enough friction developed by the rear wheels, which will then spin. Remember that friction is proportional to the normal force.

Now, no matter how stable our small car is, it will be useless if we cannot fit passengers into it. For example, an adult passenger needs at least 55 cm for shoulder space, and two passengers side by side will need an interior width of 1.25 m. Taking into account these ergonomic constraints as well as the dynamic ones mentioned above, designers have proposed the small cars shown in Fig. 8.4.

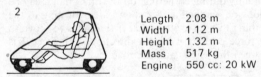

1

Length	1.78 m
Width	0.91 m
Height	1.22 m
Mass	385 kg
Engine	350 cc: 15 kW

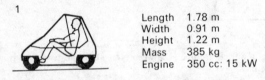

2

Length	2.08 m
Width	1.12 m
Height	1.32 m
Mass	517 kg
Engine	550 cc: 20 kW

Fig. 8.4. Town cars for one and two passengers.

But a surprising fact emerges. In spite of their small size these cars will give little saving in roadspace in the lengthwise direction. You will remember that cars have to be spaced sufficiently to allow for emergency stops. If a car stops abruptly the following car must have enough time for its brakes to dissipate its kinetic energy. Now the maximum braking force exerted by the wheels is proportional to the weight of the car, which in turn is proportional to the mass of the car. And the kinetic energy of the car ($\frac{1}{2}mv^2$) is also proportional to the mass of the car. Thus, since the mass of the car influences both the braking force and the kinetic energy, it can be seen that it does not matter whether the car is big or small; the stopping distance will not be affected. In practice there is a slight saving. For example, at a speed of 75 km/h the average conventional car would stop in 23 m, whilst a small car would require 21.5 m.

There is, however, the possibility of big savings in road widths since the small cars would only require a lane width of about 2.1 m. But it is important to note that there is little to be gained by mixing small cars with other traffic. Gains in road use can be achieved only by setting aside one or two lanes for town cars, or by building light overhead structures to carry the small car traffic.

Of course much less parking space will be required, and this would be a decided advantage. But people are not enamoured with small unorthodox cars. If their use is to increase, steps must be taken to make it clear to the public that there are advantages, and this would require the provision of segregated ways and some form of government financial incentive to prospective purchasers. One possible incentive would be to ban all cars, except the small ones, from the city centre.

Another possibility would be to ban *all* cars from the city centre and to provide a rent-a-car system for small cars. The Toyota Company have developed a 'Town Spider' system which uses two-passenger vehicles operating from a network of parking lots. The driver activates his vehicle by inserting a magnetically treated card into an electronic card receiver. Time charges and other data are transmitted automatically to a central computer through an on-board transmitter/receiver using weak radio waves.

The city of Amsterdam has also experimented with a car rental system. Vehicles were powered by electric motors, carried two passengers and could travel at 30 km/h. Keys were available at about £10 with an initial membership fee of £20.

Vehicle Propulsion

Having looked at the potential advantages of changing the shape of the car, let us now examine some recent innovations that are intended to mitigate or eliminate some of the less attractive aspects of the car.

The internal combustion engine has been criticised because of the amount of pollution it generates. In order to ensure complete combustion of the fuel, it is necessary to supply the engine with 14.6 times more air than fuel, by weight. If the fuel is not burnt properly, pollutants will be exhausted to the atmosphere, leading to problems such as those experienced in Los Angeles, where in 1963 cars exhausted daily 2.5 million kilograms of hydrocarbon, 8 million kilograms of carbon monoxide and 1 million kilograms of nitrogen oxide. Unfortunately, the fuel is rarely burnt properly. When an engine idles, as it mostly does in town traffic, the mixture is too rich, i.e. too much fuel, and carbon monoxide and unburnt hydrocarbons such as hexane are exhausted. At higher speeds, such as on motorways, the mixture is weak and oxides of nitrogen are formed. These are all harmful to humans. As a result steps have now been taken to reduce exhaust pollution mainly by using afterburners which burn the exhaust gases before they are emitted from the exhaust pipe.

Jet engines and gas turbines burn cheap fuel and cause little pollution, and these are now being proposed for use in buses and trucks.

Concern over pollution and over petrol shortage has caused a revival of interest in steam engines. I suppose, for most of us, mention of the steam engine conjures up visions of red-hot coals, large boilers and clouds of smoke. But in the US, Mr William Lear, a multimillionaire, has been a leader in the rebirth of the steam engine. He has convinced many people that steam can replace petrol in the future car, and has developed a boiler and a steam turbine engine for a motorcar. It weighs about the same as a conventional internal combustion engine and develops about 250 kW. The steam engine is remarkably silent in operation, due to the fact that it does not rely on the explosive burning of petrol. It is practically pollution-free into the bargain.

Electric power provides another alternative to the internal combustion engine. Electric vehicles need no petrol. Their power can come from coal or nuclear fuel, both of which have much greater reserves than oil (see Chapter 10). Indeed, they could be powered from solar power, which is there for the taking.

The power can be supplied to the vehicle in two ways: (*a*) externally by means of rails or of overhead wires; (*b*) internally by means of batteries. Now rails and wires restrict a vehicle's freedom of movement, and as far as the private vehicle is concerned (or a delivery vehicle) the use of a battery is a more attractive proposition.

Batteries

Battery-powered vehicles are not new. In the early days of petrol vehicles, electric cars were formidable rivals. Indeed, an electric car, La Jamais Contente, held the world's speed record of 106 km/h in 1899. Compare that with today's milk float that trundles about at 8 km/h. But the racing car had one critical disadvantage—it destroyed its batteries in one brisk surge. Today's engineers are searching for a battery that will allow a car to trundle *and* to have a reasonable top speed.

The **lead-acid battery** is by far the most commonly used today, but it is very heavy so that its use is restricted mainly to delivery vans and to material transporters in factories. For example, the UK Electricity Council found that electrification of a Mini Traveller needed batteries with a mass of 300 kg. The vehicle was capable of a top speed of 70 km/h, had a range of 48 km on level ground, and could accelerate to 35 km/h in 5 seconds. The low energy density is the main obstacle to the more widespread use of the battery in vehicles. The lead-acid battery, for example, has an energy density of about 5.5 watt-hours per kilogram. Of those available today, the **zinc-air battery**, with an energy density of about four times this value, looks promising, but unfortunately it requires ancillary equipment such as an electrolyte pump and an air compressor. And quite recently, attention has been focused on the **sodium-sulphur battery** whose energy density is slightly better than that of the zinc-air battery. Although it requires an operating temperature of 350°C, its use in delivery vehicles looks extremely promising.

The other main disadvantage of batteries is their long recharging time, which can be from 5 to 10 hours; hence their use entails short journeys and recharging facilities at the destination. Thus, on the whole, and in spite of the fact that some considerable research has been devoted to both batteries and vehicles, the future of the battery-operated car is very doubtful. There is hope, however, that city centre delivery vehicles will come to make greater use of battery power.

There was hope too that the **fuel cell**, such as those used in Apollo space vehicles, would be a much better means of supplying electric power for traction. In the fuel cell (Fig. 8.5), unlike the battery, fuel is fed in continuously, the reaction products are continuously removed, and the cells remain practically unaltered. Hence no recharging period is required. In many ways our own bodies are a type of fuel cell; the food in our blood, which is an electrolyte, is oxidised catalytically by enzymes to produce energy, some of which is electrical.

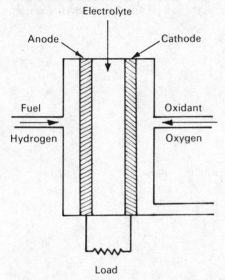

Fig. 8.5. The fuel cell converts chemical energy directly to electrical energy.

The fuel cell is a direct energy-conversion device and could conceivably have an efficiency as high as 90 per cent since thermal processes are not involved. (We saw in Chapter 1 some of the hazards involved in locking energy up in the form of heat.) In theory the hydrogen fuel cell could provide 25 times greater energy density than the lead-acid battery, but prospects of producing an economically viable cell for a car are not bright.

Linear Motors

The basic method of converting electrical to mechanical power is well known (Fig. 8.6(*a*)). Knowing that the force on a current-carrying wire is mutually perpendicular to the magnetic flux and to the current, we can imagine a propulsive system in which the current-carrying wire was borne along by the vehicle and the track included a continuous series of magnets. The vehicle would still need wheels to support its weight and to guide it. A more practical arrangement would carry the electromagnets with the vehicle, and current would be directed by moving contacts through a plate fixed to the earth (Fig. 8.6(*b*)). Greater thrust could be achieved by putting a number of these units in series.

Moving contacts can, however, be unreliable, and they can be eliminated if the induction motor principle is used. In this a rotating magnetic field is generated by a set of two or more coils supplied with two or more phases of alternating current. The coils are fixed in the stator or primary (Fig. 8.7). If the moving rotor or secondary is a conductor, such as an aluminium tube, eddy currents are induced in it by the rotating magnetic field and these in turn cause the rotor to rotate in the same direction as the field. Such a motor can be 'unwrapped' into a linear structure in which the rotor is now only a flat plate (see Fig. 8.7). This is the new popular **linear induction motor.**

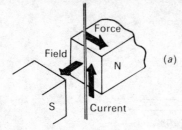

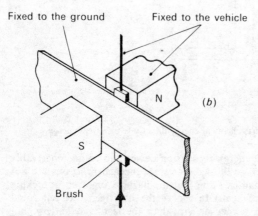

Fig. 8.6. (*a*) The force on a current-carrying wire is mutually perpendicular to the magnetic flux and to the current. (*b*) If the wire and the magnets are attached to a vehicle it can be driven along by the generated force.

The most attractive configuration from a purely technical viewpoint is to have the primary embedded continuously along the vehicle's track and to suspend the secondary, which could be merely a sheet of aluminium or copper, from the underside of the vehicle. However, this would be very costly, the long fixed stator windings along the track being very expensive. The alternative of laying the secondary on the track and carrying the primary on board, also has its disadvantages, for in order to get power to the primary it is necessary to collect it from a wayside live rail, and at high speeds and high power levels this is difficult. Of course, power could be

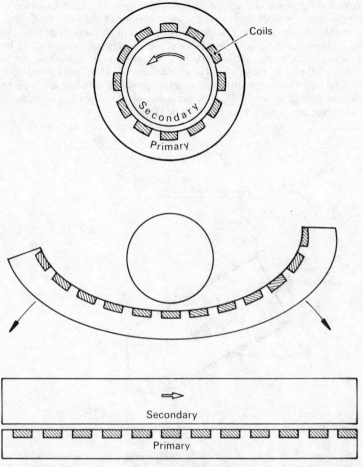

Fig. 8.7. A linear motor results from the unwrapping of a conventional induction motor.

generated on board the vehicle, but this could be noisy and uneconomical, especially if one is thinking of railway systems requiring perhaps 1.5 MW per vehicle. Another problem with the linear induction motor is that it requires the maintenance of a gap of about 1 cm between the primary and the aluminium rail. In spite of these problems engineers are confident that the linear motor will make a growing impact on transportation systems. We shall refer to it again at a later stage.

Pneumatic Propulsion

Pneumatic propulsion offers another possible means of driving a vehicle. One system uses the pressure of the atmosphere to drive the vehicle along a tube at high speed and indeed such a scheme was proposed to solve the transportation problems of the north-east corridor of the US—the densely populated strip along the east coast from Boston through New York to Washington. An evacuated tube is used, and the train is pushed along by admitting atmospheric air behind it. If the train is 3 m in diameter, a propulsive thrust of about 80 kN can be developed. This, aided by running the tube downhill for a distance, enables speeds of 600–800 km/h to be achieved with ease. Braking can be accomplished by supplying air in front of the train and by using an uphill slope (Fig. 8.8).

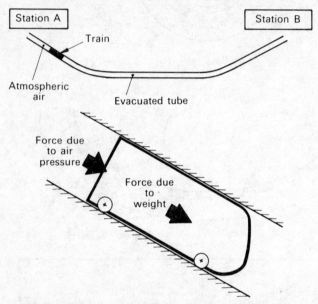

Fig. 8.8. A train can be sucked along a tunnel at 800 km/h.

Of course there is nothing new in engineering, and it is interesting to note that Brunel's 'atmospheric railway' of 1844 used the principle of pneumatic propulsion. A 35 cm diameter pipe between the rails was exhausted of air by pumping stations along the line, and a close-fitting piston in the pipe was attached to the train through a continuous slot in the top of the pipe. The slot was made airtight by a leather flap which was opened automatically by the passage of the train and immediately sealed.

Suspension and Support

This is another area where technology can offer improvements in design. The steel wheel running on a steel rail has many advantages, not the least of which being the low friction forces involved, e.g. only 1.5 kN of thrust is needed to maintain the motion of an engine weighing 1.5 MN. However, at very high speeds great precision is required in the manufacture of wheel and track, and in the alignment of curves. The use of the wheel also intro-

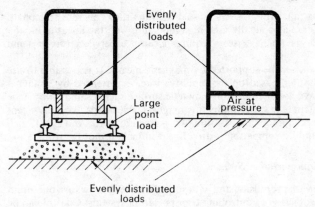

Fig. 8.9. Use of an air cushion allows a more even distribution of load.

duces loading problems. For example, the load in a railway carriage is more or less uniformly distributed over the floor, but this must be concentrated through the frame, chassis and axles to an area of contact between rail and wheel of about the size of a penny (Fig. 8.9). This, of course, gives rise to very high stresses at contact. The load must then be redistributed by means of rails and sleepers and ballast, so that the intensity is reduced to a value which the ground can support. Now this variation in loading, from low intensity to high intensity back to low intensity, can be avoided if an air cushion is used for support. The air cushion can distribute the load over a wide area of track, thereby avoiding the constructional expense associated with the use of railed vehicles. The elimination of wheels and conventional suspension allows the frontal area, and hence the aerodynamic drag of the vehicle, to be reduced by almost one third. And weight is also reduced by a similar factor.

Since there is no contact between the vehicle and the track, there is negligible friction and no wear of the guideway, and there is also the possi-

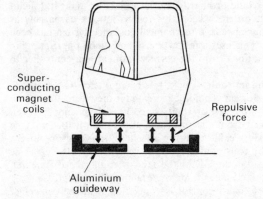

Fig. 8.10. Magnetic levitation using superconducting magnet coils.

bility of using fairly rough surfaces such as unfinished concrete or asphalt. Such systems are also ideally used in conjunction with the linear induction motor, which as we have seen requires clearance between primary and secondary.

Vehicles can also be supported by magnetic means. It is possible to use **repulsive levitation**, based on the repulsive force experienced by a magnet moving above a conducting, non-magnetic surface such as aluminium. One proposed system uses superconducting magnets which carry large persistent currents if kept at a low temperature (4 K)—see Fig. 8.10. And again, such a system would be compatible with a linear induction motor drive.

Control of Transportation Systems

Having dealt with propulsion and with suspension, let us now turn our attention to the problems of controlling transportation systems. Control can be divided into two major subsections: (*a*) **micro control** and (*b*) **macro control.**

Micro control is concerned with individual vehicles, their guidance, their speed and their headway. Macro control, on the other hand, is concerned with the operational aspects of the transportation system as a whole and ensures, for example, that vehicles are routed by the quickest path to their destination, that the optimum number of vehicles are using the system at a given time, etc.

Micro Control

Let us start with micro control, which can itself be divided into further subsections: first **guidance and steering.**

Trains running on rails are examples of guideway systems in which the vehicle is physically constrained to follow a particular path. The bus, like the train, is a fixed route system, but unlike the train, the driver has the responsibility of guiding the vehicle along its route. We all know from personal experience that the movement of buses in towns is greatly hindered by motorcar traffic, and vice versa. A possible solution to this problem is to run the buses on reserved tracks, or 'bus lanes', on those sections of the route where congestion would ordinarily restrict speed and make it difficult to predict schedules. In order to keep this reserved track as narrow as possible and to allow high speeds, a mechanical guidance system has been proposed (Fig. 8.11(*a*)). This uses a follower arm and guide pin running in a slot in the track. Deviations from the correct path cause a controller to take corrective action.

The motorcar is a random route vehicle, this being one of its big advantages, but road capacity for cars could be increased if they were automatically guided on certain stretches of road, since lateral spacing between vehicles could be reduced from the present 0.9 m to about 0.4 m. One proposal to this end uses a cable carrying an alternating electric current which can be detected by coils mounted on the vehicle (Fig. 8.11(*b*)). This system has been developed in the UK by the Transport and Road Research Laboratory, and has allowed a car to drive at 60km/h 'hands off'.

Another subsection of micro control is concerned with the **control of velocity.**

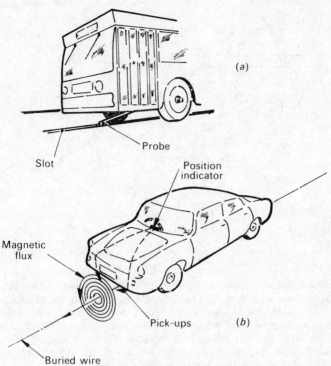

Fig. 8.11. Vehicle guidance systems: (*a*) bus guided by a probe in a slot; (*b*) car guided by buried cable.

In railways, an exact knowledge of speed is essential so that schedules can be adhered to, and so that safety can be assured. The basis of a proposed new British system of train control is a zigzag conductor laid between the tracks (Fig. 8.12). Such a system has potential for automatic urban transport systems. As the train passes over this 'wiggly wire', the train's pick-up coil generates a signal. This is amplitude-modulated at a frequency depending upon the wavelength of the wire and the speed of the train. For example, at 160 km/h and a zigzag wavelength of 3.7 m, the modulation frequency is 12 Hz (160 000 ÷ 3600 ÷ 3.7). Changes in this modulation frequency can be used by a controller to correct the speed.

This system can also be used to pass binary information to the train. In this case, coils or loops are placed in the wire; these induce discrete transient signals in the train's pick-up coil, the sense of the signal depending on the way in which the coil is wound. Fig. 8.12 shows how the binary signal 101 can be generated. Such signals could inform the driver of changes in gradient, for example.

In the present-day motorcar, speed is measured by a speedometer which is connected directly by a cable to the car's drive shaft. But in future many vehicles will not use drive shafts or wheels, and they will require alternative

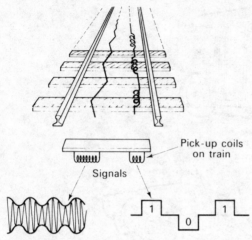

Fig. 8.12. Measuring the speed of a train. The frequency of the modulated signal is proportional to speed.

means of speed measurement. In addition to the 'wiggly-wire' system just described, another possibility is to bury small metal plates in the guideway at fixed intervals. A magnetic switch carried in the vehicle could sense the presence of these plates and would generate a pulse each time a plate is passed. The time between pulses is a measure of vehicle velocity. This system would also give position information by adding the number of pulses and multiplying by the distance between plates.

An anonymous reviewer of this book told me that speed measurement by pulses reminded him of his youth when one could calculate the speed of a train by counting the number of 'diddley-dums' in 41 seconds; this gave the speed in miles per hour. In those days the rails were laid in standard 60-foot lengths with slight gaps between to allow for thermal expansion, and as the wheels rode over these gaps they made a 'diddley-dum' noise. You should be able to prove that the number of pulses or 'diddley-dums' in 41 seconds is in fact equal to the speed in miles per hour.

The third and last area of micro control is concerned with the **control of headway** between vehicles.

If we wish to use the maximum capacity of a route, then clearly we would use as many vehicles as possible and make them travel as fast as possible. However, safety considerations dictate that a minimum distance must exist between vehicles at a given speed. This distance, or **headway**, consists of two parts:

(*a*) The distance covered by the vehicle during the reaction time of the driver and the control system; this is proportional to the velocity.

(*b*) The distance covered by the vehicle as it stops after the control system has applied the brakes; this varies with the square of the velocity (the kinetic energy of the car).

Hence the total road space occupied by a vehicle is the sum of the vehicle's length and this safe distance or headway. Thus, as speed increases the car needs more road. But the capacity of a route is measured in units of vehicles per hour. There is thus a conflict: at low speeds cars are packed tightly together on a given length of road, but since they are going slowly the actual number of cars per hour passing a point is low. On the other hand, at high speeds the distance between cars is large, and although they are moving quickly we get the same result that the actual number of cars per hour passing a point is low. There is an intermediate speed at which the route capacity is a maximum. For example, if traffic consists of cars each 3.6 m long, the optimum speed is 25 km/h, giving a route capacity of about 2000 cars per hour.

With conventional car traffic, drivers usually keep headways which are less than safe, but in future automatic vehicle systems this could not be tolerated. One proposed method of measuring the distance between vehicles uses microwave transmission of information (Fig. 8.13). It requires that each vehicle knows its own position and that it can communicate directly with the vehicle behind it. As mentioned earlier, position can be determined by counting the pulses generated by a magnetic switch as it passes over plates in the guideway. This information is then relayed to the following vehicle by means of a microwave transmitter. The computer in the second vehicle then calculates the difference in the positions of the vehicles and initiates corrective action if necessary.

Macro Control

As an example of overall control of a transportation system we shall consider the problem of routing traffic automatically through a network of roads or guideways, so that travel time is minimised. This problem was introduced in Chapter 6, and a plan of the system was given in Fig. 6.10. One-way traffic was assumed and there were 16 stations. We have already seen how dynamic programming can be used to determine the shortest routes between stations, and in Chapter 6 this particular problem was solved assuming two-way traffic. It is left to the reader to determine how the restriction to one-way traffic affects the solution.

For a large system, a central computer (Fig. 8.14) would identify these optimum routes, for even in this simple example this task is a considerable one. The central computer can then compile a steering matrix which lists the instructions to be followed by the steering system when on a given line, heading for a given exit or destination. Part of the steering matrix is shown

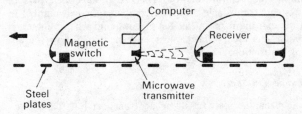

Fig. 8.13. Control of headway. Each vehicle knows its own position from the pulse count. The first vehicle transmits its position to the second vehicle which then adjusts the headway to a safe value.

in Fig. 8.14. (The full matrix would have 24 columns.) For example, the matrix indicates that if a vehicle is on line FG and its destination is M, then it should turn left at the next intersection.

When determining the optimum routes and hence the steering matrix, the central computer can also take into account such things as diurnal variations in traffic density and emergency situations. The central computer would then relay the information to local computers, one at each station. For example, station G, which has routes GH and KG running into it, needs the information in the GH and KG columns of the matrix.

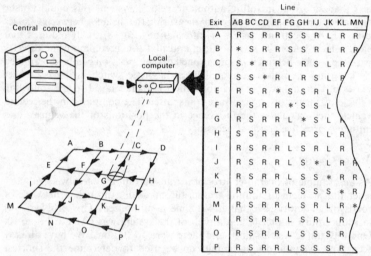

Exit	AB	BC	CD	EF	FG	GH	IJ	JK	KL	MN
A	R	S	R	R	S	S	R	L	R	R
B	*	S	R	R	S	S	R	L	R	R
C	S	*	R	R	L	R	S	L	R	
D	S	S	*	R	L	R	S	L		R
E	R	S	R	*	S	S	R	L		
F	R	S	R	R	*	S	S	L		
G	R	S	R	R	L	*	S	L		
H	S	S	R	R	L	R	S	L	R	
I	R	S	R	R	L	S	R	L	R	
J	R	S	R	R	L	S	*	L	R	R
K	R	S	R	R	L	S	S	*	R	R
L	R	S	R	R	L	S	S	S	*	R
M	R	S	R	R	L	S	R	L	R	*
N	R	S	R	R	L	S	R	L	R	
O	R	S	R	R	L	S	S	S	R	
P	R	S	R	R	L	S	S	S	R	

Fig. 8.14. Macro control. A central computer and local computers at each intersection direct vehicles according to the matrix shown.

This information in turn must be transmitted to the small guidance computer on board the vehicle. At the start of a trip, the passenger indicates his destination by setting a register on his vehicle with an exit number. As the vehicle approaches an intersection, the local computer transmits the appropriate column guidance information. This, as we have seen, lists all the exits in the system. The vehicle computer compares the exit number in its register with the exit numbers in this list of data, and when a match is found the computer sets up the mechanical equipment to follow the appropriate steering instructions.

New Ideas for Public Transport

The bus has long been the stalwart of urban transport but it has been caught in a downward spiral over recent years. Increasing car ownership has reduced the demand for public transport, and bus operators, subjected to financial constraints, were forced to reduce service levels and to increase fares. This in turn encouraged a greater shift to private cars, and technical developments in buses have not been sufficient to stop this swing. Can we provide a more attractive form of public transport?

Personal Rapid Transit

We have today the technology to allow us to build a completely revolutionary form of public transport system, which would have most of the car's advantages and few of its disadvantages. These systems are known as personal rapid transit systems, or sometimes as automatic taxi systems. The taxi is the only personal transit system which is in significant use today. It is routed individually for the passenger to his selected destination, and every large city has a thriving taxi service in spite of the fact that it costs much more than a bus.

Personal rapid transit (PRT) systems use taxis operated automatically on a network of guideways such as those discussed earlier. The vehicles will be small, carrying 2–6 passengers, and will be electrically propelled. Stations will be on side tracks off the main line so that, unlike buses and trains, only those passengers wishing to disembark at a given destination need stop at that destination (see Fig. 8.15). The by-pass station has the advantage that vehicles would wait for people, rather than people for vehicles. The use of automatic control, as discussed earlier, allows much shorter headways between the vehicles and the choice of shorter travel paths, thereby increasing route capacity.

The use of small vehicles has two main advantages. First it allows privacy

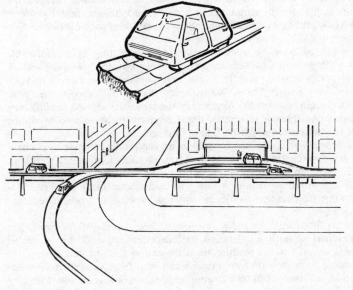

Fig. 8.15. Personal rapid transit.

for the passenger—all he has to do is punch his destination on a keyboard and he will be carried there non-stop. Secondly, the use of small light vehicles permits the use of inexpensive guideways; this is extremely important since the cost of the guideway so dominates the cost of the total system that it is better to use many small cars than a few large ones. In addition lightweight structures can be more pleasing aesthetically, and stand a better chance of blending in with the surrounding architecture.

Moving Pavements

Moving pavements offer exciting possibilities for public transport. For mass transport over relatively short distances, the moving pavement presents some considerable advantages. Top speeds of 30 km/h are being considered, and this would give a **capacity** of 30 000 passengers per hour for a pavement of width slightly greater than a metre. This is ten times greater than the capacity of a motorway lane!

But how do we get passengers on to a pavement moving at 30 km/h? There are several ways of doing this, and perhaps the most obvious means is to use several pavements running in parallel at different speeds (Fig. 8.16(*a*)). Indeed, such a system was used a long time ago in Paris at the Great Exhibition of 1900, and it was such a success that it inspired H. G. Wells to refer to it in *The Sleeper Wakes*: 'Each [platform] moved to the right, each perceptibly slower than the one above it, but the difference in pace was small enough to permit anyone to step from any platform to the one adjacent, and so walk uninterruptedly from the swiftest to the motionless middle way.'

However, in spite of Wells' optimism, the discontinuities between the various stages present major safety hazards; probably sufficient to deter Aunt Maud! What is needed is a means of accelerating passengers smoothly from rest up to the maximum pavement speed.

This can be done by using a moving belt consisting of a deformable mesh, analogous to a garden trellis (Fig. 8.16(*b*)). In the low-speed section the belt width expands slightly, so decreasing its length and hence its velocity. In the high-speed zone the width of the belt is forced to decrease; hence its length and velocity increases in the same manner as a fluid passing through a constriction. Obviously the mesh needs to be covered by sliding plates to stop passengers from falling through!

A rotating disc could also be used as an accelerator if passengers could walk from its centre to its edge. Laboratory studies suggest a safe limit of 0.22 rad/s on the rotational speed, so even to attain an edge speed of only 10 km/h would require a disc with a diameter of 26 m. When disembarking, deceleration would be achieved by walking from the edge to the centre.

Many other ingenious mechanisms have been proposed for producing the required smooth acceleration and deceleration, and the interested reader is referred to the reading list at the end of this chapter.

Now we have seen that automation can help to solve the urban transportation problem, let us turn to another area where automation is making a striking impact.

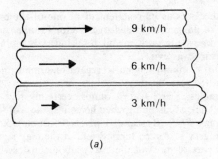

(a)

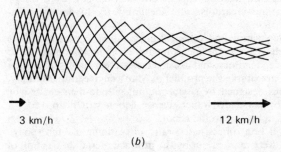

3 km/h   12 km/h

(b)

Fig. 8.16. Moving pavements: (*a*) parallel sections; (*b*) deformable mesh.

8.3 Robots

Mankind has always been facinated by the concept of the robot slave, built in his own image and responsive to his every whim. From earliest times he has been infatuated with his own shape and his living functions, for the most miraculous mechanism which has ever existed on Earth is the human body itself. The efforts of creative man to make copies of himself indicated the very beginnings of technology.

There was also the desire to emulate the act of creation, and this desire was expressed in art and literature. The earliest prehistoric men have bequeathed the cave paintings which were their first attempts to produce facsimiles of themselves. In literature it was the poet Homer who perhaps first envisioned robots. We can find reference to them in Book 8 of the Iliad where Homer refers to '. . . (maids) made of gold exactly like living girls; they have sense in their heads, they can speak and use their muscles; they can spin and weave and do their work: . . .'—in short, they were robots.

Homer's robots were benign, but in later times the robot developed a sinister image. Mary Shelley's classic horror story, *Frankenstein*, was published in 1818, and it had a strong impact on the way in which society would regard robots and other man-made men for decades to come. One reason for the story's success was its restatement of one of the enduring fears of mankind—that of dangerous knowledge. Victor Frankenstein was another Faust, seeking knowledge not meant for Man. His ambition to create a soulless intelligence was evil.

From Frankenstein on, mankind tended to regard man-made creatures as inherently evil, bound to turn on their creators, or at the very least to cause them endless troubles. Karel Čapek almost certainly had Frankenstein's monster in the back of his mind when he coined the word 'robot' in 1920, in his famous play *RUR* (Rossum's Universal Robots). *Robota* is the Czech word for drudgery or forced labour, and in the play robots are created by Man to take care of most of the heavy labour; instead, true to the ghost of Frankenstein's monster, they turn on their creators. The pattern was set, and science fiction writers from then on portrayed robots as malignant, unnatural creatures, who inevitably try to destroy mankind.

Fortunately, Isaac Asimov had sufficient imagination to portray robots as benign beings. He rejected the idea that the creation of robots places Man in God-forbidden territory, and insisted instead that robots, like other mechanical devices, are simply the product of Man's engineering ingenuity. Robots are machines designed by engineers, not pseudo-men created by blasphemers. Asimov persuaded other science fiction writers to adopt a more magnanimous approach to the robot.

Asimov's science fiction robots had 'brains of platinum-iridium sponge and the brain paths were marked out by the production and destruction of positrons'. To design the 'positronic' brains of these robots required a huge and intricate new branch of technology that Asimov called 'robotics', a word that is now fully accepted by modern technologists who use it to describe the study, design and manufacture of robots.

When he was 19 years old, Asimov wrote a story called *Runaround*, in which he introduced his now famous laws of robotics:

1. A robot may not injure a human being or, through inaction, allow a human being to come to harm.

2. A robot must obey the orders given it by human beings except where such orders would conflict with the first law.

3. A robot must protect its own existence as long as such protection does not conflict with the first or second law.

These laws contain ambiguities, but it is a remarkable fact that this original fictional foundation can still provide a useful design guide for the cybernetic engineer.

Automata

Everything that we have discussed so far has been of a fictional nature. What about Man's attempts to build actual working replicas of himself? These were called automata, or machines acting by means of concealed mechanisms. I suppose the earliest automata were those constructed by Hero of Alexandria about 100 BC, powered by water and steam. But the heyday of the automaton did not really arrive until the eighteenth century. The makers of these automata were the Mozarts of the mechanical world, and it is impossible to overestimate the ingenuity and delicacy of their work. Let us look at a few examples.

Henri Louis Jaquet Droz (1752–91) had artistic and musical talents as well as mechanical expertise. He assisted in the creation of three famous automata, now at Neuchâtel, which were first shown to the public over 200 years ago in 1774. These automata consist of a child writer, a child draughtsman and a young woman playing the organ. They possessed an uncanny reality and were for this reason called androids, since they were imitations of actual life-size characters, rather than artificial figurines.

The Writer is quite small like a curly-headed four-year-old, and he holds a goose quill in his right hand, whilst his eyes follow the tracing of his words. When the mechanism is started, the boy dips his pen into an inkwell, shakes it twice, puts his hand at the top of the page and begins to write. The letters are formed carefully, using both thin and thick strokes, with gaps between the words, and at the end of the line the hand moves lower down the paper to commence a new one.

About the same period another mechanical wizard, Baron von Kempelen (1734–1804), was greatly renowned. Although producing many ingenious devices, including a talking device that could produce all five vowels, he is perhaps best remembered for a confidence trick. This was his notorious mechanical chess player, which on a famous occasion played and beat Napoleon (Fig. 8.17). Although this had the appearance of being fully automatic, and fooled people for a long time, it has since been proved to have been a trick automaton worked by a hidden legless accomplice. In spite of this, it was an impressive piece of mechanical ingenuity. It had to be wound up and then it actually picked up the pieces and placed them with its left hand. When the automaton checked its opponent it nodded its head three times. The hidden accomplice must have been a good player for chess is difficult enough without having to play it upside down and back to front.

We can see that the automaton was intended to entertain, and it was usually designed to do one thing and one thing only. On the other hand,

Fig. 8.17. Von Kempelen's mechanical chess player.

the robot, as we know it today, is not designed for amusement but for a job of work. It is not restricted to one task, but can be programmed to do different work cycles. And, of great importance, the robot can make decisions and can behave in an intelligent manner.

Instructing and Controlling the Robot

How do we tell a robot or an automaton what we want it to do? How do we control it? We saw earlier in Chapter 4 how a camshaft could be used to control the motion of an actuator, and how the timing of the sequence of movements of several actuators could be determined by such means. For example, an elementary industrial robot, called a **'pick and place'** device, requires three actuators—one to grip a component, one to lift it and one to transfer it. A camshaft with three cams would be used for this purpose. More complicated robots might use punched cards or tapes, plug boards and drum memories.

The camshaft controller is basically a device for controlling the timing of a sequence of events. But timing is only one of many things that may have to be controlled during the operation of a robot. Consider, for example, the device we have just discussed. How would we go about adjusting the distance through which the component was to be moved? One way would be to have adjustable mechanical stops which limited the motion of the actuators, but such a method would be feasible if the changes in distance were required only occasionally. If accurate and variable positioning is required, such as the control of aircraft elevators, we would need to use feedback control (Chapter 4).

I have already emphasised that if we wish to control something we must be able to measure it, and we saw in Chapter 3 that technology has now devised a great number of instruments for measurement of variables such as speed, temperature, humidity and so on. For our robot arm we could use a potentiometer to determine position. Another variable of particular interest in robotics is force, particularly in relation to the design of the robot's hand. It would be important, for example when handling fragile objects, that the robot hand would not cause damage. This can be avoided by including a force control system, in which the forces in the robot's fingers would be measured by electrical strain gauges.

Industrial Robots

Let us now have a look at some of the industrial robots that are becoming more and more common in manufacturing industry. There are now over 100 manufacturers of industrial robots in the world. Japan is reputed to have more than 13 000 in action in its industries, whilst the UK employs only 1100 at the time of writing. During the whole of this century there has been a growing interest in industrial automation, for if the working actions of a human being can be reduced to a succession of classified movements, then there is no fundamental reason why those stereotyped movements should not be made by a machine rather than by a human operator.

A typical industrial robot is shown in Fig. 8.18. It consists of a body that can rotate about a vertical axis, an upper arm that can rotate about the shoulder, a forearm that turns about the elbow, and a wrist assembly that can twist, move up and down and left and right. Various special-purpose hands can be attached to this wrist. These can be used for gripping, for lifting with suction pads, for drilling, for welding, for paint spraying and so on.

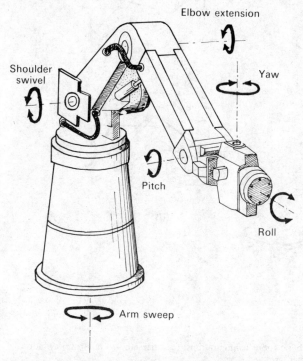

Fig. 8.18. A typical industrial robot (Cincinnati Milatron).

Figure 8.19 shows four basic robot configurations that allow the arm to move an object to an arbitrary point within a working space. They are (*a*) **cartesian**, (*b*) **polar**, (*c*) **cylindrical** and (*d*) **jointed arm**, like the robot in Fig. 8.18.

In cartesian coordinates movement is allowed along three mutually perpendicular axes. When polar coordinates are used the turret is free to rotate about both horizontal and vertical axes, while the arm is allowed to move in and out. With cylindrical coordinates the arm is allowed an up and down motion and an in and out motion, whilst the mounting turret can rotate about a vertical axis. The jointed arm configuration allows both parts of the arm to rotate independently about horizontal axes, whilst the whole structure moves about a vertical axis. In all cases the ideal wrist movement would have three degrees of freedom, rotation about the axis of the arm and about the other two perpendicular axes.

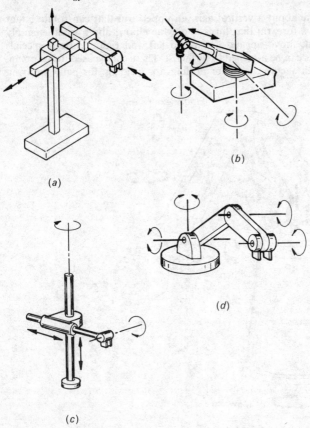

Fig. 8.19. A few basic robot configurations: (*a*) cartesian; (*b*) polar; (*c*) cylindrical; (*d*) jointed arm.

A robot's control system has two basic functions: (*a*) to get the movements of hand, wrist and arm in the right sequence; (*b*) to ensure that hand, wrist and arm are positioned accurately. There is a third basic function: (*c*) to store the space coordinates that the robot or its power actuators are to occupy during each sequence of operation, i.e. the instructions for (*a*) and (*b*).

One of the most interesting features of industrial robots is the ease with which they can be given this information. One method requires the engineer or 'teacher' to drive the robot through each step of its program, using a portable programming unit. After each discrete movement the teacher registers the position on his hand-held unit and this information is stored, step by step, in an electronic memory. Thereafter the robot relies on this memory to tell it what to do, and it will repeat those instructions over and over again until it is reprogrammed.

Another interesting form of learning is used by some robot manufacturers and is particularly useful in those cases where the robot's hand has to move through a continuous path. Powder and paint spraying are examples of such cases. Here the teacher, who would be a skilled sprayer, would manually guide the robot's hand through the full spraying sequence. This is stored in a microprocessor memory system and allows the robot to duplicate the exact motion of the skilled operator.

Industrial robots are ideal for jobs that are dangerous or unpleasant, and we could list many such jobs—heat treatment, handling of glass, spot welding, furnace tapping, handling corrosive substances, and so on. And the robot can be used for boring, tedious work, such as loading components into machines or on to conveyor belts.

But these robots, although being tireless and immune to extreme environments, are not terribly bright. A child could learn more quickly. And they are stupid. If a piece of metal is presented to the robot in a position that is different from the one it expects, it will be none the wiser and will continue to treat this piece as though it were the same as all the others. More recent robots, however, have been given some sense: they have been equipped with sensors so that they can detect what is happening in their vicinity. Touch and vision are two important examples. Tactile sensors, using strain gauge techniques (see Chapter 3) are being used to allow robots to determine if they are gripping objects correctly. And vision sensors, relying on television, let the robots know if components are absent or incorrectly orientated. The provision of these sensors improves the 'intelligence' of the robot. But can a robot think and learn? Many research laboratories are now working in the area of artificial intelligence.

Robot Intelligence

Alan Turing, one of the pioneers of the philosophy of computing, proposed what has become known as Turing's Interrogation Game as a definition of artificial intelligence.

The idea behind this definition is that of the party game in which the player, by having a conversation with a person, has to determine whether that person is a man or a woman. A third person acts as an intermediary so that the player cannot see or hear, but gets all the information by asking questions. Turing suggested that if a similar game were set up with a machine and the player could not tell whether the hidden 'opponent' was a human being or a machine, then it could be asserted that the machine had achieved the level of intelligence of the human being. This is really saying that if a machine appears to be intelligent then we may as well assume that it is intelligent.

Let us look at a few things we associate with an intelligent being. First the ability to understand and to converse in a natural language would be a necessary attribute. Much research is being carried out in this field. For example, researchers at MIT have devised a program called 'Doctor', which allows a computer to carry on an apparently intelligent conversation with a patient in a hospital. Essentially all the computer does is to pick out a key word and fit its response around it. Here is an extract from a 'conversation' in which the computer is playing the role of the psychiatrist.

Computer:	'How do you do. I am the Doctor. Please sit down at the typewriter and tell me about your problem.'
Patient:	'I have lost my baggage.'
Computer:	'You say you have lost your baggage.'
Patient:	'I feel lost without it.'
Computer:	'Do you often feel lost without it?'
Patient:	'No.'
Computer:	'Why not?'
Patient:	'Because I normally do not lose things.'
Computer:	'What other reasons might there be?'
Patient:	'My mother trained me not to lose things.'
Computer:	'Tell me more about your family.'
Patient:	'I have lost my baggage!'
Computer:	'Did your parents ever lose you?'

You can see that the computer's responses are innocuous, to say the least, but the surprising thing is that 60 per cent of the patients in a Massachusetts Hospital who talked to 'Doctor' insisted that they were conversing with a real person. Thus we could say that the computer has passed Turing's test. This is a simple example, but even so it can produce what passes for intelligence. Much more advanced work is being carried out in this field.

What other things would we expect an intelligent machine to be capable of? I suppose we would be disappointed if it couldn't play games—noughts and crosses at least! But there is a basic difference between asking a computer to find the square root of a number and asking it to play a game. The method of solution to the square root problem can be defined exactly, but in the case of a game such as chess, things are not just so clear-cut. The human player relies on his memory of the outcome of moves in earlier games, on hunches and on his ability to compare and evaluate the strength of different moves. There is no clearly defined solution—or is there? Surprisingly the answer is Yes! Algorithms can be written for chess.

How do they work? When it is the computer's turn to play, it could consider all the legal moves open to it and all of the board configurations that would result. Typically, about 30 moves are possible in the middle game. For each of the board configurations resulting from these moves, all the moves the opponent might make in reply have to be considered. This would result in 30 times 30, or 900 board configurations, and so on. If this procedure were to be followed to every possible termination of the game, then it would be possible for the computer to select the best possible move at every stage, and hence to become the world champion. However, it has been estimated that, even using a very fast computer, the number of years it would take to decide on a single move in chess by this method would be represented by a one followed by 90 zeros. In other words, this exhaustive exploration of every possible move is quite impossible as a strategy for playing chess (Fig. 8.20).

Nevertheless, existing programs for chess depend on a partial exploration of possible moves and the resultant responses by the opponent. These may look up to four moves ahead and when the computer has determined all the possible board configurations four moves ahead (about a million) it

Fig. 8.20. A chess-playing computer that looks at every possible move.

assigns a score to each position. The score or value of each configuration would take account of such things as the number of pieces held by each player, the safety of the king, the control of the centre, the promotion of pieces and so on. The computer then favours the configuration with the highest score and makes its move towards that end.

This indeed is not unlike the way a human chess player tackles the game, and of course it is the human programmer that teaches the machine the tricks of the trade. But machines are now being developed that think for themselves! They can remember past games and learn from them; they can adjust their method of evaluating a position as the game proceeds.

There are many other things that we would expect an intelligent machine to be capable of. The ability to recognise patterns would be essential. When we read a book, for example, we have to recognise symbols and groups of symbols. In everyday life we have to recognise from their visual images all sorts of things—cars, lamp-posts, people. In addition, the recognition of sound patterns is necessary for the understanding of speech. These tasks are easy for people but are very difficult to perform by machines. But they must be mastered if mobile robots are to develop. Again a lot of work is being done in this area, and several advances have been made such as fingerprint classification, reading printed documents, putting components in the right position for assembly, analysis of particle tracks in bubble chambers, and so on.

I suppose, too, that we would expect an intelligent robot to be able to learn. We mentioned earlier that some computer programs for chess are able to learn from the success or failure of earlier moves. Recent research has produced robots capable of learning to find their way through a maze. A few years ago the grand finals of the Great Electronic Mouse Race were held in New York. The challenge was to design a robot mouse capable of finding its way through a maze in the shortest possible time. In principle there was nothing to stop anyone entering a real mouse, except for the rule 'The deposition of any material substances on the race course is prohibited'.

What emerged were tiny robots powered by batteries and steered by beams of infrared light which detected the walls of the maze and altered course accordingly. Learning was provided by microprocessors so that the robots could improve their performance over several runs, memorising the correct turns to make. The winner improved its time of 1 minute 41 seconds for the first run to 31 seconds for its third.

Learning systems of this kind could have many applications; for example, a vacuum cleaner or a polisher controlled in this way could be used to clean large buildings such as airports. They could be turned on during the night and left to determine the room's boundaries and to clean everywhere in between.

8.4   Conclusions

Let me conclude by distinguishing among three different aspects of robot development. First there is the hope that a scientific study of robots will throw light on the nature of intelligence and how the brain works. This will put robots in the centre of some of the most exciting technological developments in the coming years. Secondly, there is the use of robots to do jobs that humans cannot do, or prefer not to do. We have already mentioned some of these, and there will be many more exciting ones such as space and undersea exploration. Such robots are of obvious benefit to mankind. Finally, the third aspect of robots is crucial. There is a fear that robots will replace humans everywhere and there will be mass unemployment. This we shall discuss in Chapter 10.

8.5   Exercises

1. List ten examples of different technologies that you can identify within a 100 metre radius of your home.

2. Write a 500-word essay on the history and development of the bicycle.

3. (*a*) A car brakes when travelling downhill. Modify Fig. 8.1 to account for the slope and show how this affects the tipping stability. (*b*) Modify Fig. 8.2 to account for camber and show how this affects stability during a turn.

4. Write a 500-word essay on the potential application of the steam engine to road traffic.

5. List and compare five methods of propelling a vehicle.

6. (*a*) Distinguish between macro control and micro control of a transport system. (*b*) Give an example of micro control in a rail system and in a road system. (*c*) Give an example of macro control in a rail system and in a road system.

7. Write a 500-word essay on developments in personal rapid transit.

8. What is a robot? How does it differ from an automaton? Is a radio-controlled aircraft a robot? Are traffic lights robots? Is a juke-box a robot? Is a Jacquard loom a robot?

9. (*a*) Describe two methods of programming a robot. (*b*) Describe two methods of programming an automaton.

10. What are the basic features of an industrial robot? Describe the four basic robot configurations.

11. List 12 applications where the use of robots is advantageous.

12. (*a*) What is machine intelligence? (*b*) Describe how a computer might be programmed to play tic-tac-toe (noughts and crosses).

8.6 Further Reading

Aldiss, B., *Science Fiction Art*, New English Library, 1975.
Ardley, N., *The Amazing World of Machines*, Angus & Robertson, 1977.
Baxter, R. and Burke, J., *Tomorrow's World*, BBC, 1970.
Bendixson, T., *Instead of Cars*, Penguin, 1977.
Black, I., Gillie, R., Henderson, R., and Thomas, T., *Advanced Urban Transport*, Saxon House Lexington Books, 1975.
Computers and the Year 2000, NCC Publications, 1972.
Gunston, B., *Transport Technology*, Geoffrey Chapman, 1972.
Hillier, M., *Automata and Mechanical Toys*, Jupiter Books, 1976.
Marsh, P., *The Robot Age*, Abacus, 1982.
Persall, R., *Collecting Mechanical Antiques*, David & Charles, 1973.
Rorvik, D., *As man becomes machine*, Abacus, 1975.
Spencer, D. D., *Game-playing with Computers*, Hayden Book Co., 1975.
The Future of Work, Open University, T262/10–11, 1975.
The Robots are Coming, NCC Publications, 1974.
Thring, M., *Machines—Masters or Slaves of Man?*, Peter Perigrinus Ltd, 1974.
Thring, M., *Man, Machines and Tomorrow*, Routledge & Kegan Paul, 1973.
Traffic in Towns, HMSO, 1963.
Young, J. F., *Robotics*, Butterworths, 1973.

9
Industry: Structures and Operation

9.1 Introduction

Industry is the manifestation of technology. In this chapter we shall identify the different sectors of industry, and special emphasis will be given to manufacturing industry in its role as a producer of wealth. The structure of a typical company will be examined and the relationship between management and unions will be explored.

9.2 Industry

In Chapter 7, on the methodology of technology, we saw that the first important step in the process was the identification of a need or want. The satisfaction of that need or want is the objective of technology, and industry is the means of meeting that objective. Most of our wants can be classified under the headings of either **goods** or **services**. Goods are tangible things such as pencils, cornflakes, shoes, motorcars and so on. Services are less tangible. We want to have our rubbish bins emptied; we want to have our children educated; we want to be entertained; we want to be flown to Florida for our holidays, and so on. In order to live a comfortable life we seek both goods and services, and the provision of these goods and services is known by economists as **production**.

Direct and Indirect Production

There are two methods of production: direct and indirect production. When we provide our own goods and services we are involved in direct production. When we cooperate with others in the provision of our and their goods and services, we are involved in indirect production.

A little reflection will show that direct production is inefficient. Imagine being a castaway on a desert island. You have to satisfy your basic needs of food, water, clothing and shelter; and you have to do it all by yourself. How would you feed yourself? You could hunt animals, fish, lay traps or collect berries and fruit. Clearly this would take up a lot of your time. What about water? You might be lucky to find a convenient stream. On the other hand, you may have to dig a well, or you may have to collect rainwater, or you may have to collect condensate; again all very time-consuming. And what about clothing? You could skin animals and use their hides or furs, or you might attempt to weave coarse materials from plant fibres. But again this is a time-consuming task. So you can see that direct production to meet your own needs is a full-time job. There would be little, if any, free time to allow you to do anything that was not related to your basic needs. You would remain poor and undeveloped as primitive Man did for thousands of years.

Indirect production is much more efficient. Imagine how different it would be if you had been stranded along with three colleagues. You would find that working together gives better results, mainly because each of you could specialise. The best hunter could specialise in finding food; the best with his hands could specialise in making pots and clothes; another could specialise in providing shelter and cooking. Each person could spend all of his time on his particular specialisation, and this could provide enough to meet not only his own needs in relation to his own specialisation, but those of his colleagues also. And even better, if surplus food were produced, then, in the happy event of making contact with another group on another island, you could begin to trade. You could, for example, trade your surplus of pots for their surplus of bows and arrows. There may even be some time free to allow some of your colleagues to concentrate on providing services rather than goods. For example, someone may take on the role of the witch doctor! So life begins to become more comfortable and meaningful.

The Division of Labour

Specialisation is central to modern society. We have teachers, bricklayers, actors, doctors, hairdressers, engineers—all specialists involved in indirect production, providing each other with goods and services. Adam Smith, the great British economist, concluded that wealth came from such specialisation, and in his book *The Wealth of Nations* he illustrated this by reference to the manufacture of pins. In 1756 he wrote:

'A man not educated to the business could scarcely perhaps with his utmost industry make one pin a day, and certainly could not make twenty. . . . But in the way this trade is carried on it is divided into a number of branches, of which the greater part are likewise peculiar trades. One man draws out the wire, another straightens it, a third cuts it, a fourth points it, a fifth grinds it at the top for receiving the head; to make the head requires two or three distinct operations; to put it on is a peculiar business, to whiten the pins is another; it is even a trade by itself to put them into the paper. . . .'

Smith calculated that such a division of the labour into ten parts, using one man for each part, would result in the production of 48 000 per day, or 4 800 per man per day. One man doing all of the operations would produce a maximum of 20 pins in a day. (It is interesting to note that a modern pin-making machine produces about ten pins per second.) Henry Ford also recognised the advantages of specialisation when he developed the motorcar industry from a system in which one man built a complete car, to one where hundreds of men, each performing one operation repeatedly on an assembly line, produced an endless stream of cars.

There are many advantages in specialisation: people can choose the work they like; time for learning the necessary skill is reduced; there is greater use of equipment, and so on. But against these there are several disadvantages which have led to a reassessment of the division of labour. These are: monotony causes strain and fatigue; the worker's initiative and motivation are checked; an interruption in one section of the production line affects the whole; and so on.

What do Producers need?

We have defined production as the provision of goods and services. No matter what the form of production, the producer will have four basic requirements: (*a*) premises; (*b*) workers; (*c*) machines and tools; and (*d*) materials.

(*a*) The premises could be a factory, an office, a garage, a shop or a back room.

(*b*) The workers could be welders, engineers, teachers, clerks, secretaries, bus drivers, postmen, magicians, etc.

(*c*) The machinery and tools could include motors, furnaces, saws, hammers, scissors, pens, magicians' wands, and so on.

(*d*) The materials are not only those used in the actual product, but also those required to assist the production. These could include wood, plastics, cloth, paper, stone, metal, etc., and in some cases raw materials such as trees, oil, iron ore or coal, might be required.

The Three Types of Production

Production falls into three categories: primary, secondary and tertiary production. **Primary production** is the first basic stage of production. It exploits the Earth's natural resources in three ways:

1. **Extraction:** this covers mining, quarrying and fishing—three activities that extract raw materials directly from the Earth's surface. These include the many metallic ores and coal, oil, gas, diamonds, gold, platinum, sand, gravel, etc. We shall see in the next chapter that these extracted resources, including fish, may run out very quickly if we are not careful.

2. **Cultivation:** this covers agriculture and forestry. Carrots, wheat, tea, beans, tobacco, chicory, rhubarb, apples and timber are typical products.

3. **Animal husbandry:** the rearing of cattle, sheep, pigs, poultry, deer, mink; for their meat, milk, hide, furs. Nowadays fish husbandry, or fish farming, is becoming increasingly popular too.

Some examples of occupations in the primary sector of production are: coalminer, farmer, fisherman, fur trapper, goldminer, herdsman, oil driller, pearl diver, whaler.

Secondary production alters the primary products so as to form manufactured articles. The manufacturing industries and the building and construction industries are involved in secondary production. A manufacturing industry can produce capital goods or consumer goods. **Capital goods** are those products which will be used by other producers, e.g. nuts and bolts, wire, nails, machinery. **Consumer goods** are those that are ready to go to their final destination—to the consumer, to you and me, to the man on the street. Tins of beans, electric toasters, toothbrushes, trousers and books are examples of consumer goods.

In addition to capital goods and consumer goods, the secondary sector of production covers building and construction which involves the fastening together of manufactured components to produce houses, factories, sewage plants, etc.

Typical occupations in the secondary sector are: aeronautical engineer, builder, carpenter, civil engineer, decorator, draftsman, electrical engineer, fitter, mechanical engineer, production engineer, potter, shipbuilder, tailor, technician. The secondary sector is a very important wealth producer and we shall examine it more closely later.

The third type of production, **tertiary production**, is concerned with the production of **services**. These break down into **commercial services** and **personal services**. The commercial services include the transport of manufactured goods to the customer; they provide publicity; they arrange loans, they provide telephones and letter boxes, etc. All of these are essential to the success of secondary production.

What about personal services? Think through a typical day and try to identify the number of occupations that provide you with a personal service: the bus driver, the paper boy, the hairdresser, the milkman?

Many occupations are classified as belonging to the tertiary sector. For the provision of commercial services we have clerks, accountants, bankers, captains, exporters, insurance agents, stockholders, train drivers, entertainers, psychologists, undertakers, watch repairers and so on.

It is important to recognise that a service is only of value if it has a beneficial effect on the availability of goods, or if it contributes to the general contentment of the community. We should also recognise that many of the activities in the tertiary sector rely on the wealth produced in the primary and secondary sectors of industry—education is one example. Therefore we should have a closer look at, for example, British industry to see if we can identify the important aspects of wealth creation.

9.3 British Industry

The population of the UK is about 56 million. Of these about 23 million are employed and 3 million are unemployed. The remaining 30 million comprise housewives, children and retired people. Table 9.1 shows the distribution of occupations in the UK for the year 1981.

Thus, for 1981, about 33 per cent of employees were in the secondary sector, about 3 per cent in the primary sector and about 64 per cent in the tertiary sector. So the service industries are the dominant ones—and their dominance is growing.

Fig. 9.1 shows how the distribution of UK employees has varied between 1840 and 1980. The most striking features are the continuing decrease in

Table 9.1 Distribution of occupations in the UK (1981)

	Industry	Employees (*thousands*)
Primary	Agriculture, forestry, fishing	360
	Mining, quarrying	332
Secondary	Manufacturing	6 038
	Construction	1 132
Tertiary	Gas, electricity, water	340
	Transport, communication	1 440
	Distribution trades	2 635
	Insurance, banking, finance	1 233
	Professional and scientific services	3 695
	Catering, hotels, etc.	872
	Miscellaneous services	1 542
	National government service	615
	Local government service	964
	Subtotal of employees	21 198
	Employers and self-employed	1 856
	HM Forces	334
	Total	23 388 (thousands)
	Total available workforce	26 068 000
	Unemployed	2 681 000

employment in the primary industries, the continuing increase in service industries, and the marked decline in manufacturing and construction over the last 20 years. This pattern of change is not unique to the UK; similar trends exist in the USA, Germany and Japan.

The reduction in the numbers employed in the primary industries has been caused largely by developments in technology; compare today's agricultural equipment with that of a century ago. The swing from manufacturing to service industries is also partly explained by the impact of new technology, and we shall look more closely at this in Chapter 10.

It is important to emphasise that the manufacturing industries create a lot of wealth. In the UK this sector accounts for more than one half of total exports, and these are used to buy food and other resources that the country cannot entirely provide on its own. Let us now examine manufacturing industry more closely to see how it generates this wealth.

9.4 Manufacturing Industry

The major objectives of an industrial firm are (i) to produce goods and (ii) to make a profit. Nowadays **profit** is often considered to be a dirty word, one associated with the drinks cabinet in the managing director's Rolls-Royce. But this is a narrow-minded viewpoint, and we shall now attempt to demonstrate that profit is essential to the success of an enterprise.

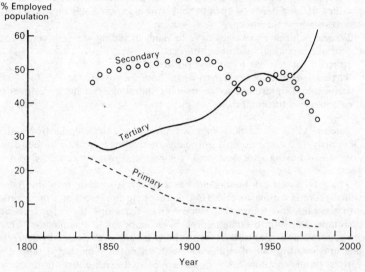

Fig. 9.1. The distribution of UK employees amongst the primary, secondary and tertiary sectors.

This requires an examination of the flow of money in an industrial company.

Our model of a company's economic system will be based on one developed by the Chemical and Allied Products Industry Training Board (CAP). First we have to appreciate that at any particular time a company has certain resources—its **total assets**—that could be turned into cash if the need arose. These fall into two categories: working capital and fixed assets. The **working capital** includes the value of the work that is in progress, the raw materials held in the stores, and the finished goods held in stock. For example, in a car factory this would include the stocks of sheet metal, tyres, gearboxes, upholstery and instrumentation, etc., as well as all those cars at various stages of completion on the production lines, and of course all the completed cars awaiting despatch in the company's car park. In addition to this hardware, working capital includes 'liquid money' which represents the difference between what the company owes at that time and what is owed to it.

The remainder of the total assets—the **fixed assets**—includes everything else that the company owns, covering chairs, catering equipment, manufacturing equipment, buildings and so on.

Knowing what the total assets are, we now have to identify the various factors that can lead to their reduction or to their increase. First, if we examine the fixed assets, it should be clear that these will tend to dwindle over the years as equipment wears out or becomes obsolete. Next we can list the company's necessary expenditure under the following headings:

Purchases: The company has to buy the raw materials that constitute its product. It also needs to spend money on electricity, gas and water and services such as transport.

Wages: Salaries and wages have to paid, including the company's contribution to national insurance and pension costs.

Interest: Interest has to be paid on loans.

Taxes: Corporation tax has to be paid to the government.

Dividends: Interest has to be paid to shareholders who have invested their money in the company.

Against all of these outgoings we have to set the income from sales. Obviously, if income exceeds outgoings the company will be in the happy position of having an excess, and it is essential that this excess is put to good use.

The total assets can be modelled as a reservoir of cash, into and out of which there is a flow of cash (Fig. 9.2). There are two flows or incomes into the tank: one represents income from sales and the other, **retained cash flow**, will soon become clearer. What happens to this income? There are also two flows out of the tank: a direct leak of cash caused by **depreciation** and wearing out of equipment, and an outlet pipe that is tapped at various points to allow the various above listed expenditures to be drawn off. The first, and often the major expenditure, is the purchase of goods and services. The difference between this expenditure and the income is known as the **added value**, since it represents the value that the company has added to the raw materials and services it has bought.

Ideally this added value should be large enough so that the cash flows to wages, interest, tax and dividends do not drain the system dry. The excess, the nasty profit we referred to earlier, is called the retained cash flow. This

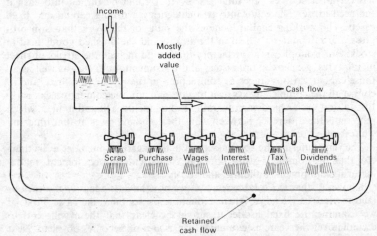

Fig. 9.2. A model of the cash flow in a company (after the Chemical and Allied Products Training Board).

can be returned to the tank where it can be used for several purposes. It could, for example, make the tank larger if it were used to buy new equipment, i.e. increase the fixed assets. It could, on the other hand, be reduced to zero by increasing wages or by paying larger dividends to the **shareholders**—those people who have bought a share of the company and expect a reasonable return on their investment. Again, it could be diverted from the company's tank into other ventures or new companies.

9.5 Company Performance

It is instructive to compare the economic performance of companies from different countries. I have chosen the motor industry, the companies being General Motors (USA), Toyota (Japan) and British Leyland (UK). The data in Table 9.2 have been extracted from the annual reports and accounts of the companies by Dr F. E. Jones, the leading protagonist of the concept of added value.

Table 9.2. Economic performance of selected companies

	General Motors (1977)	*Toyota* (1975)	*British Leyland* (1975)
Employees	797 000	44,616	164 354
Working capital	£9714m	£1215m	£430m
Capital/employee	£12 188	£27 232	£2616
Fixed assets	£4317m	£410m	£262m
Fixed assets/employee	£5416	£9190	£1592
Total assets	£14 031m	£1625m	£692m
Total assets/employee	£17 604	£36 422	£4208
Sales/employee	£36 295	£60 847	£9175
Bought-in goods and services/ employee	£17 081	£47 490	£5997
Added value/employee	£19 214	£13 357	£3178
Added value per £ of Capital	£1.58	£0.49	£1.21
Added value per £ of fixed assets	£3.55	£1.45	£2.00
Added value per £ of total assets	£1.09	£0.37	£0.75

Several points arise from this table. First we can see that the value added by each employee was much less in the British company—less than one sixth of the American employee and about one quarter of the Japanese employee. Comparing the whole of the manufacturing industries with those of Japan it has been shown that the average added value per employee for 1977 was £4893 in the UK and £13 142 in Japan.

This is critical, for added value is a measure of **productivity** and we could infer that the Japanese workforce is used roughly four times as effectively as the British. We then have to ask ourselves is this due to poor management in the UK, or to slackness on the part of the British worker?

This is a complicated issue, one that is at the heart of every industrial dispute.

Against this, Table 9.2 reveals that British Leyland produces a greater return of added value on its total assets (£0.75) than Toyota (£0.37). Indeed, a comparison of the whole manufacturing industries of the two countries for 1977 shows that the amount of value added per £ of total assets is £0.67 in the UK and £0.24 in Japan. Looked at in this way, we could conclude that the UK does much better on its more limited assets, and since we use our assets more effectively we could expect to improve our position if we had more assets.

It is interesting to calculate what increase in assets would be required to bring us up to the same provision as Japan. In 1977 the assets per employee in the UK amounted to £7330; in Japan the equivalent figure was £55 464. The difference per employee is £48 134, and multiplying this by the number of employees in the UK manufacturing industry (about seven million, Table 9.1) gives a grand total of about £350 000 million. Where could this vast amount of money come from? Should government provide it? If so, where would they get it from? Would they divert it from another budget heading such as education? Would they increase taxes to provide the money? Would you be happy to pay higher taxes? And so on. Another solution is to create more added value and to feed more of this added value back into the industry. But you will recognise that this is a circular argument—in order to make industry better, you have to make industry better!

Added value has been described as the pot of gold that keeps industry going. Let us examine how it is distributed in General Motors, Toyota and British Leyland. The breakdown is given in Table 9.3.

Table 9.3. Distribution and added value in selected companies

	General Motors (1977)		Toyota (1975)		British Leyland (1975)	
	£	%	£	%	£	%
Value added/employee	19 214		13 357		3 178	
Wages and salaries	10 084	(52.5)	4 228	(32.1)	2 933	(92.3)
Welfare and pensions	1 935	(10.1)	334	(2.5)	287	(9.0)
Rent and hire	100	(0.5)	0	(0)	0	(0)
Financial costs	186	(0.9)	187	(1.4)	198	(6.2)
Local rates	1 195	(6.2)	668	(5.0)	70	(2.2)
Corporation tax	1 938	(10.1)	2 377	(17.8)	−131	(−4.1)
Depreciation	1 572	(8.2)	3 219	(24.1)	258	(8.1)
Net profit	2 204	(11.5)	2 284	(17.1)	−437	(−13.7)

The meanings of most of the subheadings should be reasonably clear. Wages and salaries pay for labour, but it is worth noting that the pay that the employee takes home is less than this figure; income tax and social service contributions reduce it by about 23 per cent. Financial costs include interest on loans from banks. Local rates have to be paid to local councils for the use of land and local services. Corporation tax is the government's share of the profits.

Several facts emerge from Table 9.3:

(*a*) Although the wages of the British Leyland employee were the smallest, they represented the largest percentage of the added value of any of the companies.

(*b*) The total sum paid to the government—income tax, corporation tax, national insurance, etc.—was high for the American and Japanese companies. This wealth helps to keep the service industries in action. It can provide hospital beds, policemen, schools and so on.

(*c*) The large added values in General Motors and Toyota allowed a substantial sum for the replacement of worn out and outdated equipment (depreciation).

(*d*) The larger profits in the foreign companies allowed money to be reinvested in the companies. This could be spent on new techniques and equipment, both of which will tend to improve the company's performance even more.

What conclusions can we draw from these statistics? Clearly the British company was faring poorly in comparison to its foreign competitors. Its small added value did not provide enough to meet all the outflows we identified in Fig. 9.2. So salaries were low, and it was not possible to reinvest in the company.

Why did British Leyland do so badly? We have to ask the following questions:

1. Was the equipment up to it? In other words, bearing in mind our earlier argument that the total assets in the UK could be improved, had enough been spent on the latest technology?
2. Was the manpower up to it? Did management ensure the most effective use of its resources? Did employees put in a good day's work? Were staff up to date on the latest technology?

These are vexed issues and are continually under debate. We cannot hope to resolve them here, but at least you should now have a clearer picture of the importance of manufacturing industry, and how essential it is to generate sufficient added value to allow satisfactory wages to be paid, to allow for reinvestment, and to allow for payments to government for the maintenance of the service industries.

9.6   The Subsections of Industry

Although technology is central to our manufacturing industries, it needs the support of many other disciplines. It is one thing producing articles, but they have to be sold and this requires sales expertise; manpower has to be employed and this requires expertise in handling personnel—and so on. Let us see if we can identify the major important subsections of a company within the secondary sector of production.

Production

The production function is the most important: everything else depends upon it. It can take several different forms in our factories:

(*a*) **Mass production**, as the name implies, is concerned with the manufacture of goods in large numbers, such as razor blades or buttons.

(*b*) Other goods, not in such great demand, may be manufactured in a **batch production** system in which a limited number or a batch is produced. Aircraft, furniture and books are produced in this way. The extreme form of batch production is a 'one-off' production, sometimes known as 'jobbing'. This produces a single product and is often found in new engineering firms who are anxious to try their hands at anything in order to establish a reputation.

(*c*) At the other extreme we have **flow production**, a process in which raw materials enter continually at one end of a system, and emerge continuously as a product at the other end. Beer, oil, sugar and liquid chemicals are produced in this way.

There are many different types of people involved in production, including machine operators, craftsmen, technicians and technologists. In order to give you an idea how they interact it would be useful to imagine the production facility in a factory that makes golf trolleys. The workshop would need lathes for turning the wheel axles, stamping machines for cutting bracket shapes from sheets of metal, grinders for smoothing rough edges, machines for drilling holes, welding equipment for joining the pieces together and so on. In order to do these jobs there would be a need for (*a*) turners to operate the lathes, (*b*) machine operators to control the stamping machines, (*c*) welders, (*d*) fitters to assemble the various components. In addition, technicians could be needed to finalise the working drawings, to test the quality of the various pieces and of the finished article. The design of the trolley, its components and any special features needed for its manufacture, would be the responsibility of a mechanical engineer. A production engineer would ensure that the production unit was being used in the most efficient way, e.g. machines being in use as often as possible, production of axles matching production of brackets, materials in store being adequate for the demand, etc. To back all this up there would also be a need for cleaners to keep the workshop tidy, labourers to move heavy equipment about, and storemen to look after the stores.

Since it is central to the whole operation of the company, the production unit has a heavy responsibility. It must produce golf trolleys at the required rate and of an adequate standard. It must be able to meet sudden increases in demand and in a reasonable time, for the company will suffer if customers have to wait. Its staff must always be looking for cheaper and better methods of manufacture.

Research and Development

Research and development (R and D) falls into three categories:

(*a*) **Basic research** (sometimes called pure research) concerned with achieving a better understanding of an area of science that is relevant to the company's work. For example, large manufacturers of glass carry out a lot of research into the ways that glass melts and fractures, and manufacturers of microprocessors spend a lot of money on studying the science of semiconductor materials. These researchers hope to discover new materials and processes that can be developed into useful products.

(*b*) **Applied research**, aimed at demonstrating the feasibility of applying scientific knowledge, and usually involving small-scale tests of ideas. For example, knowledge about the behaviour of materials at extremely low temperatures could suggest a new type of computer memory element, and the applied research worker would test out various ways of applying this scientific knowledge to the proposed practical application.

(*c*) **Development** that gradually scales up the laboratory process to the full-scale manufacturing plant. This usually includes all the work needed to turn the researcher's ideas into something that can be made profitably. Development work is very practical, and it is worth noting that in most companies, except for the very largest, development predominates in their R and D departments; basic research plays only a small role.

The R and D department looks for entirely new products, for better ways of making existing products, and for ways of improving existing products. It has to face many tough problems. For example, in a company that makes ploughs, the R and D department could be asked to determine if an automatic control system using microprocessors would give better response to a variety of soil conditions. Again, a company that makes food containers might ask its R and D department to investigate the possibility of mildew growing on their plastic boxes if they were exported to the tropics. Finally, suppose that market surveys indicate the need for a more convenient and more accurate method of measuring blood pressure. This is the sort of problem that would be tackled by the R and D department in a biomedical engineering company. It would require the use of the methodology laid down in Chapter 7 of this book.

Marketing

We have seen that R and D turns ideas into practical realities—realities that the production department has to produce efficiently. But products are of no commercial use if no one wishes to buy them. There has to be a market for a company's goods, and the Marketing Department has the responsibility of assuring the company that a market exists before production of a new or modified article commences. The Marketing Department has to be aware of the habits, attitudes, likes and dislikes of people. As we saw in Chapter 7, there is no point in producing helmets with guns on top if people are not prepared to buy them!

The Marketing Department gathers its information through **market research**. People are interviewed, questionnaires are completed, statistics are examined, psychologists are consulted—all to determine what people will consider to be a want. Marketing experts will watch trends in taste, for example, and they will keep a wary eye on their company's competitors. They have to plan ahead. They will know how to advertise effectively.

We can get a better feeling for the work of the Marketing Department by imagining a few problems it might have to face:

1. A company manufactures baby food and wishes to expand its business by preparing foods for very old people. What information is needed?
2. A company manufactures pies using synthetic flavours. Food chemists invent a new formula that uses real fruit; it will be twice as nice but will put 50 per cent on to the cost of the pie. What to do?
3. A company manufacturing lathes wishes to advertise its product. One advertising agency says that advertisements ought to be witty; another says they ought to be serious. What to do?

Sales

Together the departments identified so far can generate an idea, convert it to reality, determine if it will have a market, and manufacture the product. Selling is the way of getting the product to the customer, and it is so critical that many companies recognise that a steady flow of orders can be assured only if people are employed whose only job is selling. These sales representatives have to be thoroughly knowledgeable about their product, and have to be prepared to roam the country, even the world, in an effort to sell it. They have many problems. For example, a sales representative of a ball-bearing company calls on one of his biggest customers to be told that he is considering the purchase of another company's bearings since they are 10 per cent cheaper. What to do? Again, a favoured customer of a shirt factory asks for an additional 10 000 shirts next week: if he gets them some of the factory's smaller customers will go short. What to do?

Finance

Earlier in this chapter we examined the flow of money in an industrial company, and we were able to sketch a model of the process in Fig. 9.2. Money is a key resource, and its management is the responsibility of the Finance Department (or Accounts Department). Accurate records have to be kept of all incoming and outgoings monies, and at the end of each year the Finance Department has to produce an account of the year's business. This is checked by an independent auditor and is available for public inspection. The compilation of this account can be a mammoth task, especially in a large company. It requires very precise monitoring of cash flow, including purchases of premises and equipment, wages and salaries, cash from customers, taxes, national insurance payments, interest to banks, dividends to shareholders.

In addition to this housekeeping role, the Finance Department may have to raise millions of pounds for a new factory, and special long-term bank loans may have to be negotiated, or a decision may be taken to issue more shares on the stock market.

As well as this, the Finance Department will have to be able to calculate the cost of manufacturing the product. If several products are being manufactured it should be able to advise on the mix of the products and the timing of their manufacture to produce the maximum profit (see linear programming in Chapter 6).

The sort of problems faced by the Finance Department are illustrated by the following examples:

(a) A company making electric toasters finds that its main competitor can make theirs for 10 per cent less. How do they do it?

(b) A company exports a lot of typewriters. They find that a new competitor is offering their German customers very good credit terms (two years to pay). What should the company do?

(c) A company manufactures paint brushes. They wish to know the financial advantage of introducing automation.

Personnel

Industry employs a great number of people with different skills and different levels of skills. The Personnel Department has the responsibility of ensuring that there are enough staff available and that they are well cared for. Ensuring that there are enough people available requires careful planning. For example, the Personnel Department has to keep a close watch on ages, so that if 10 per cent of staff are going to retire in two years' time a recruitment campaign for their replacements can be started in sufficient time. In addition to maintaining the number of staff, it is also necessary to update their skills, and this is particularly important now that technology is moving so rapidly. This requires the Personnel Department to organise training courses of various kinds.

Caring for staff includes the provision of wages, of insurance, of security, of a healthy and safe environment, and all of these factors will involve the Personnel Department in negotiations with the trade unions.

Some typical problems could be:

(a) A shirt-manufacturing company has been asked by its biggest customer to double its production over the summer months. It is necessary to recruit 30 stitchers for a six-month period.

(b) A member of the clerical staff wishes to retire early. She wants advice on the best way to do it.

(c) A welder has been dismissed for continual lateness. He has claimed wrongful dismissal and the case has to go to an industrial tribunal. What evidence has to be prepared to defend the company's action?

Industrial Relations

Every company has both workers and managers, and it is important that relations between the two are as good as possible. Both have the common aim of making the company a success, but there is a basic conflict of interest that has to be recognised. Managers are concerned about the efficient use of manpower, materials, machines and money. Workers are worried about their wages, their hours of work, the length of their holidays, their job security and their working environment. Can you see how conflicts can arise? An increase in the workers' wages is an increase in the manager's costs. If a manager introduces new machinery to increase productivity, the worker may lose his job. If the worker asks for improved working conditions, such as better ventilation or less noise, then the manager is faced with another possible increase in costs.

So conflict can arise, and in order to strengthen the individual worker's influence on discussions the **trade unions** have been formed. In the UK there are now about 12 million trade union members out of a total work force of about 26 million. There are about 400 British trade unions—groups of people with something in common such as a skill, an industry, an employer or an occupation. The largest union is the Transport and General Workers' Union (TGWU), with a membership in excess of two million. The Amalgamated Union of Engineering Workers (AUEW) is another large one with a membership of about 1.25 million. One of the smallest, the Pattern Weavers Society, has only 150 members. In 1868 the British unions came together in the Trades Union Congress (TUC) to discuss common objectives, to form a common view and to enable working people to speak with one voice.

The main objectives of trade unions are to improve their members' terms and conditions of work and to bring about an improvement in the quality of life of working people. This requires the unions to debate with both employers and government. They operate at three levels: the workplace, the industry and government. At the workplace and in industry the union uses a process called collective bargaining to negotiate terms and conditions of employment with the employer. At government level in the UK, the TUC meets with government to discuss such issues as forms and levels of taxation, economic policy, the National Health Service, etc.

There are two groups of officials who negotiate on behalf of the members of a union: union workplace representatives and full-time officials. On the factory-floor the shop steward and in the office the office representative act as union workplace representatives. These are unpaid jobs, and these people work at their own job when not involved in union business. They are elected by members, and they have the responsibility of representing members' views to the employer. The other group of officials, the full-time officials, are also elected by members. They are appointed by the union, and work for it full-time.

A shop steward's day can be hectic and varied, especially in the larger companies. Typical problems might be:

(*a*) In order to avoid possible injury to their feet, workers are expected to wear safety shoes. One of the union members refused to wear them because they were uncomfortable. He has been injured by a falling crate. Can the union help him to claim for damages?

(*b*) Management wish to use robots for welding. This could mean the loss of 20 jobs. What can the union do to protect these employees?

(*c*) A worker has been discovered drunk on the job on two occasions. The foreman intends to sack him. What should the union do?

Ancillary Services

In addition to all of the above activities a large firm may well have need of an internal system of transport, a means of distributing goods to customers, a legal department, a sports organisation, store keepers, secretaries, computing services, consultants, security staff, etc.

9.7 Company Organisation

We have now identified most of the activities to be found in a modern company. How are they coordinated in practice? How are clashes of interest between different activities resolved? This requires a company organisational structure that clearly identifies authority and responsibility at all levels. But before going further it would be worthwhile to define the words 'authority' and 'responsibility'.

According to Webster's Universal Dictionary, authority is 'power, the legal right to command and to enforce obedience'. Responsibility, on the other hand, is 'the state of being legally or morally answerable, personally accountable for action, performance of a duty, fulfilment of an obligation, etc.' So you can see that it would be rather nice to have authority without responsibility, for you could direct operations without having to care too much if they are successful or not. But it would be most unpleasant to have responsibility without authority, for you could be kicked in the backside because a particular operation failed—and you would not have the authority to make the necessary corrections. Thus it is important that company managers have both authority and responsibility.

Let us now examine how this authority and responsibility might flow throughout a manufacturing organisation. We shall choose, for illustration, a management structure typical of very large manufacturing companies like British Leyland, General Motors, etc. However, like human beings, no two organisations are exactly alike, and factors such as size and relevant technology lead to different forms of management structure. This should be borne in mind.

Fig. 9.3 shows the operating structure of a very large manufacturing company. The **board of directors** sits at the top of the tree. Members of the board are normally shareholders in the company, and are elected at the annual general meeting of the shareholders. The duties and powers of the board are laid down in the **Companies Acts 1948 to 1981.** Board members are concerned with major matters of company policy: What improvements can be made? What are we doing wrong? Do we know enough? Having reached a decision at boardroom level, the directors leave the matter in the hands of the **Managing Director.** He, the Chief Executive, bears the overall responsibility for achieving the targets set by the board. Like President Truman, the Managing Director (MD) can truthfully say 'the buck stops here!' But the MD is only human, so in a large company he has to share his task with other managers. This is called 'delegation'.

Fig. 9.3 shows a total of 36 managers; 6 executive managers and 30 department managers. For example, the R and D manager is an executive manager responsible for four departments—design, drawing, product development and research—each of these departments having its own departmental head. Each department in turn will have its own chain of authority/responsibility. For example, the head of the production en-

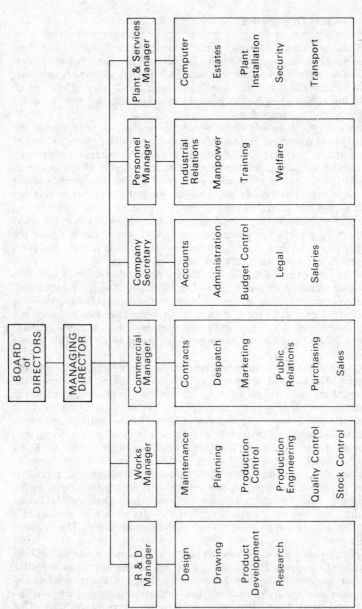

Fig. 9.3. A model of the organisation of a large company.

gineering department would be assisted by a production assistant, foreman, technicians and operatives.

You will recognise Fig. 9.3 as a model of a large company (see Chapter 6). There are several other possible models. The pyramid has the MD at the top and the machine operatives and others at the bottom. In the cartwheel model, the departments are the spokes with the MD (the boss) at the centre. A more recent model, promulgated by the disillusioned, is the mushroom model in which management keeps the workers in the dark most of the time, occasionally throwing manure on them!

The danger with all of these models is that firms appear to be organised in a very formal and inflexible way. In fact, companies are changing all the time and these models cannot cope with the many special cases and loose ends that occur. Again, the various activities are shown as discrete boxes, but in reality there is a good deal of overlap between boxes. Nowhere is this overlap more obvious or important than in the small company, where one man may have to wear many hats.

The **cooperative** presents another interesting form of company organisation. Cooperatives are groups of individuals who have recognised that they can achieve a common aim more effectively by working together. Their aims can vary from getting groceries at a lower cost to obtaining more secure jobs.

In the UK the most developed form of cooperation is seen in the **consumer cooperatives**, in which people come together to buy goods in bulk. This indeed was the basis of operation of the very first 'Co-op' founded in 1844 in Rochdale, England. Foodstuffs were bought in bulk at wholesale prices and sold to members of the cooperative at market price. Profits were split among the members in proportion to the value of their purchases. Today you can find a cooperative store in nearly every high street.

An important principle of cooperation is that of **democratic control.** In the conventional company shareholders can hold as many shares as they can afford to buy, and when important policy decisions have to be made each shareholder has as many votes as he has shares. In cooperatives each member has only one share and only one vote, so the individual is considered to be every bit as important as the money.

We are concerned with manufacturing industry in this chapter, and the consumer cooperative does not fall into that category. But the **worker cooperative** does. In these organisations the workers contribute both capital and labour. They are the owners as well as the workers and hence they will be the sole beneficiaries if results are good—and the sole losers if they are bad.

In the UK the worker cooperatives have not been nearly as successful as the consumer cooperatives. But the picture is different in countries like France, Italy and the Basque provinces of Spain. There the worker cooperatives have many successes to their credit, including well established enterprises recognised as industrial leaders in their field. The group of 70 small- to medium-sized enterprises around Mondragon in Spain provides a spectacular example of successful cooperation of this kind.

9.8   Exercises

1. (*a*) What is production? (*b*) Distinguish between direct and indirect production. (*c*) What are the advantages and disadvantages of specialisation?

2. What are the four major needs of producers? Give examples of each in relation to the production of (i) cattle, (ii) safety pins, (iii) a taxi service.

3. (*a*) What are the categories of production? (*b*) To which category does each of the following belong? Mink farmer, judge, painter, milkman, forester, surgeon, footballer, chemist, panel beater, pilot, soldier, weaver, market gardener.

4. Write 250 words explaining why service industries are expanding in the developed countries.

5. (*a*) Distinguish between working capital and fixed assets. (*b*) Identify the various expenditures incurred by a typical manufacturing company. (*c*) Why is added value so important? (*d*) Describe the uses that can be made of retained cash flow.

6. (*a*) List six major subsections or departments within a large manufacturing organisation. (*b*) Describe (50 words each) how these departments support the production function. Give a typical problem that each may have to solve.

7. (*a*) Distinguish between jobbing, batch production, mass production and flow production. Give three examples of each. (*b*) How does mass production affect (i) the product and (ii) the employee?

8. A typical manufacturing company employs a hierarchy of technical people—professional engineers, technicians, craftsmen, machine operators. Give examples of the work of each, and show how each relates to the other.

9. (*a*) What is the function of the Marketing Department? Give examples of the sorts of problems it might encounter. (*b*) A publishing company wishes to expand its business by producing Braille books for the blind. What information does it need?

10. (*a*) Why are trade unions needed? (*b*) What are the duties of a shop steward? (*c*) What is collective bargaining?

9.9   Further Reading

Curzon, L. B., *Economics for 'O' level*, Macdonald & Evans, 1981.

Holliday, L., *The Integration of the Technologies*, Hutchinson, 1980.

Industry in Close-up, Metal Box Ltd.

Industry: Men, Money and Management, Book H. 'Science in Society', Heinemann, 1981.

Mogano, M., *How to start and run your own business*, Graham & Trotman, 1982.

Pitfield, R. R., *Administration in Business Made Simple*, Heinemann, 1982.

Pitfield, R. R., *Company Practice Made Simple*, Heinemann, 1983.

Whitehead, G., *Business and Administrative Organisation Made Simple*, Heinemann, 1981.

Whitehead, G., *Economics Made Simple*, Heinemann, 1982.

10
Technology and Society

10.1 Introduction

Technology is a powerful agent for change. It can help Man achieve the ideal, comfortable, challenging and rewarding life. But if handled carelessly or with malice, it can be a great force for destruction and misery.

Technology provokes debate. When is technology good? When is it bad? On the whole, for example, one could argue that advances in medical engineering would benefit mankind, and that this is an example of 'good' technology. But it could raise more problems than it solves. Who should receive this costly treatment? How long should a patient be kept on an artificial kidney machine? And so on. On the other hand, developments in missiles could be considered to be an example of 'bad' technology. But if we stop developing our weapons will other countries follow suit? What about all the people employed in those industries? And so on.

Technological innovation often opens up moral issues, and it is important that these are discussed in an informed and objective manner. It is all very well to get steamed up when a particular issue affects us personally, but the other fellow's point of view has to be heard too. Vested interests have to be watched. How much say should the road builders have in discussions about expanding the motorway network? How much say should the oil companies have in decisions about reducing the poisonous lead content of petrol?

Then there is the other great debating point: how much blame should a technologist carry if his product is used for evil purposes? Should Nobel be blamed for inventing dynamite? Should Bell be blamed for the doubtful transactions and plots that take place on the telephone? Should Ford be blamed for deaths on the road? Man, by his very nature, is filled with wonder and curiosity about nature, the universe and the physical world. He will always attempt to solve its riddles, and in so doing will come up with new technologies. And I would contend that there is no technology that cannot be abused by those who wish to do so. Being facetious for a moment, one could use a screwdriver as an offensive weapon and one could commit suicide by jumping off the top of a chest of drawers! But let us look at a more serious example—laser technology. It can be used for microsurgery, for metalworking, for communications. Now it is being developed as an outer space weapon. Is this a bad thing? If so, is the technologist to be blamed—or the government? It seems to me that in a democracy a government has the responsibility of monitoring the directions of its technologies, ensuring always that the best for all is being achieved.

Now these are very important issues which will raise themselves quite often throughout this chapter. In order to be more specific I thought it would be helpful to look at three areas of debate: (*a*) the environment and the Earth's resources, (*b*) technology for the underdeveloped countries, and (*c*) men and machines.

10.2   The Environment and Earth's Resources

The description of Earth as a spaceship is a very useful idea, for it forces us to recognise that, just like in a space ship, we have to ensure that the environment that we astronauts have to live in is maintained in a healthy condition. Our spaceship is a finite size and carries a finite supply of essentials. We cannot therefore take too many passengers on board; nor can we use up our rations in an indiscriminate manner.

One of our big problems is that of exponential growth (we met this earlier in Chapter 5). Many of us tend to think of growth as a linear process, i.e. the quantity or size grows by the *same amount* in each unit of time. For example, if you put £10 in a box every month then your hoard of money will increase linearly. But linear growth is not a common phenomenon. Exponential growth, in which the quantity or size grows by the *same multiplying factor* in each unit of time is more common. If, instead of putting your money in a box, you were to put £100 in a bank at a compound interest rate of 10 per cent, then at the end of the first year you would have £100 × 1.1 = £110. This amount would accrue interest so that at the end of the second year you would have £110 × 1.1 = £121. Thus in the first year your savings would grow by £10, and in the second year by £11. Indeed, in about seven years, your £100 would have been doubled. A simple rule of thumb says that for reasonably small growth rates (less than 20 per cent) the **doubling period** is approximately 72 divided by the percentage growth rate.

This doubling process is an awesome one, for in any one doubling period the variable changes by an amount equal to its growth during its entire history. This is nicely illustrated by the story of the rampant water hyacinth whose area doubled every day. The owner of the pond noticed that the hyacinth covered 2 square metres one day, 4 the next and 8 the day after that. The pond had a total area of 4096 square metres but the owner decided to wait until it was half covered before he took action. Now although it would take 11 days to cover half the pond, the owner did not appreciate that it would only take one day more to cover the pond completely. Thus his corrective action would come at a rather late stage, and he would have only one day in which to act. And this is the danger of many of our responses to exponential growth—the longer we postpone action the worse the problem becomes.

Another example of the mind-boggling nature of exponential growth is given by the story of the Persian courtier who, having presented his king with a beautiful chessboard, asked in return for 1 grain of rice on the first square, 2 for the second, 4 for the third and so on. The king foolishly agreed, not knowing the power of exponential growth to generate enormous numbers. The total amount of rice required would greatly exceed the world's total production. The last square itself would require a total of 9 223 372 036 854 775 808 grains.

The Population Constraint

Compound interest is an example of a positive feedback process, in which the more we have the more we get. We discussed other examples in Chapter 5, where it was seen that positive feedback was associated with a runaway situation. Take population: as it increases, births increase and these increase the population, in turn giving rise to more births; and so on. This is offset to some extent by deaths, but the net effect is a world growth rate of about 2.1 per cent per annum (100 people per minute)—see Fig. 10.1. The present world population is about 4×10^9, and this is expected to double in about 34 years.

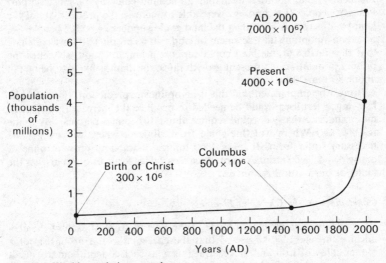

Fig. 10.1 World population growth.

Clearly, unbridled growth of population cannot continue indefinitely. There are several constraints: first the obvious one of lack of space. If mankind suddenly had an attack of claustrophobia and spread itself out as thinly as possible on the Earth, how much privacy do you think you would have? Each person would find himself about 120 metres from his nearest neighbour and this would shrink to about 60 metres in 70 years' 'time. Neanderthal Man, by way of contrast, has a possible separation of around 3.5 km.

Biologists have found that many animals evolve mechanisms for avoiding excessive population increases. Flour beetles, when crowded, generate a gas that not only kills their larvae but also acts as an antiaphrodisiac. Some animals eat their young when crowded, and others such as lemmings go mad with stress, so that even the sight of a female can give a male a heart attack. What about humans? Is there an optimum population for the Earth? Clearly, from a social point of view, this must lie somewhere between the extremes of loneliness and overcrowding. And it also depends on the amount of resources, land, fuel, etc., available, their rate of usage, pollution generation and many other effects. No sensible answer can be given without due consideration of the important interactions that take place among these factors. Let us now look at some of them.

Constraints on Agricultural Land

Agricultural land is a finite commodity. There are about 3.2×10^7 km^2 suitable for agriculture and about half of that is in use at present. Development of the remaining land would be very costly, including clearance, irrigation and fertilisation. Assuming that the present world average of 4000 m^2 would produce sufficient food for each person, then it is not hard to calculate that the total available arable land could at the most support 8×10^9 people. If technology were able to increase the productivity of the land by a factor of three, then the land could support 24×10^9 people. So, based on an optimistic assessment of land and its productivity, we conclude that the Earth's arable land could support a limit of about six times the present population. At present growth rates, this limit is likely to be met in about 80 years' time.

It must be emphasised that this is an optimistic prediction. For example, how much fertiliser would be needed to produce the increased yield? Estimates indicate that we would require about 10^9 tonnes per year. (A tonne is 1000 kg.) Where will this come from? Indeed, where will we get the necessary water from? It takes 4000 tonnes of water to grow a tonne of maize—and 15 000 tonnes to produce a car! This leads us on to discuss the limits of our natural resources.

Constraints on the Chemical Elements

Various estimates have been made of the Earth's store of each of the chemical elements. It is suspected that the core of the Earth consists of a molten alloy of iron and nickel. But this is at such a depth and at such a temperature (3700°C) that it is very unlikely that technology will be able to tap this resource like it does the oil wells.

Our ores come from the outer crust, and so far very little of this crust has been explored below a depth of 3.5 km. As far back as 1925, Clarke and Washington attempted to estimate the proportion of the chemical elements contained in the outer 16 km of the crust. They analysed over 5000 samples of rock from all over the world during their researches. More recent work, restricted to a depth of 3.5 km, came up with very similar results. These are given in Table 10.1 (see references 6 and 10 in the Further Reading).

You will note that many of the more familiar metals occur in very small percentages. It is surprising, for example, that Man ever had the luck to discover copper. But bear in mind that these are average values and the ores that Man found were much richer in copper—indeed, in some cases they consisted of nearly pure copper.

Table 10.1 Chemical elements in the Earth's crust

Element	Per cent mass	Element	Per cent mass	Element	Per cent mass
Oxygen (O)	46.60	Hydrogen (H)	1.4	Copper (Cu)	0.005 5
Silicon (Si)	27.72	Titanium (Ti)	0.44	Lead (Pb)	0.001 3
Aluminium (Al)	8.13	Manganese (Mn)	0.095	Tin (Sn)	0.000 2
Iron (Fe)	5.00	Carbon (C)	0.02	Tungsten (W)	0.000 15
Calcium (Ca)	3.63	Chromium (Cr)	0.01	Mercury (Hg)	0.000 008
Sodium (Na)	2.83	Nickel (Ni)	0.0075	Silver (Ag)	0.000 007
Magnesium (Mg)	2.09	Zinc (Zn)	0.007	Gold (Au)	0.000 000 4

The actual mass of elements in the first 3.5 km of the crust is enormous. The total mass is about 1.5×10^{18} tonnes, and of this aluminium, for example, makes up about 1.2×10^{17} tonnes. This is certainly impressive but the question is—how much of it is available? It is unlikely that we would ever be able to recover all of this material, and even if we did we would find that many of the compounds of aluminium are unsuitable for use as ores. They may not contain sufficient aluminium to justify the effort and energy of extraction, and the particular chemical combination may require a very expensive refining process. For example, garden clay is so rich in aluminium that a hundred barrow loads could make a small aeroplane. But it would take a fortune to extract the 25 per cent of aluminium from its chemical combination with oxygen, silicon, iron, calcium and magnesium. At present we depend on bauxite for our aluminium, for it is a rich ore containing about 38 per cent aluminium.

So we can see that although there are great masses of material in the Earth's crust, the actual available and useful amounts are much less. These amounts are li8ted in Table 10.2 (see reference 8) for some of the more popular metals in the outer 3.5 km of the crust.

Column 2 is determined from Table 10.1 and gives the total resource in tonnes. Column 3 gives the minimum grade of ore below which it is today considered to be uneconomical to attempt extraction. Advances in the technology of extraction and of ore processing could reduce the figures in column 3, and they could also be reduced if people were prepared to pay a

Table 10.2 Earth's available reserves of selected metals

(1) Element	(2) Resource (Tonnes)	(3) Minimum economic grade of ore (%)	(4) Percentage greater than minimum	(5) Identified reserves (tonnes)	(6) 1975 Annual Demand (tonnes)
Lead	2.0×10^{13}	5.0	4.0	1.3×10^8	3.9×10^6
Nickel	1.1×10^{14}	1.0	1.5	8.4×10^7	7.03×10^5
Copper	8×10^{13}	0.6	0.4	3.1×10^8	7.4×10^6
Iron	8×10^{16}	30.0	25.0	3.5×10^{11}	4.7×10^8
Aluminium	1.2×10^{17}	38.0	23.0	3.0×10^9	1.16×10^7

lot more for their metals. Column 4 gives the percentage of rock containing a grade of ore greater than today's minimum economic one. Column 5 lists the amounts of metal obtainable from those rocks having a satisfactory grade of ore. Note that this column is headed 'reserves'. A **resource** is the quantity of material that is potentially available for human purposes; a **reserve** is that part of a resource that is known to be exploitable. Finally, column 6 lists the 1975 annual world demand for these metals.

Constraints on Fuels

Later we shall attempt to estimate how long these metal reserves will last, but before doing that let us take a look at some other very important resources—our fuels, **coal** and **oil** in particular.

The fossil fuels make a very important contribution to our growing energy needs (Fig. 10.2). At present each citizen of the USA uses about 250 kWh every day of his life, whereas in India the figure is nearer 8 kWh.

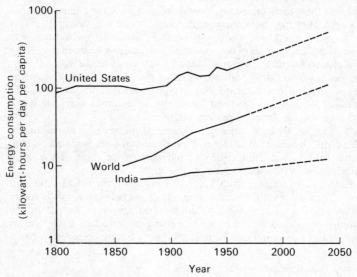

Fig. 10.2. Use of fossil fuels.

Coal is a sedimentary rock formed from the remains of plant life. It occurs in stratified deposits, usually continuous over wide areas and frequently cropping out at the surface. The older a coal the greater the 'coalification' or change from the original vegetable matter. This proceeds through peat, brown coal and lignite, hard bituminous coal to anthracite. Some of the United Nations' statistics (references 6 and 18) on indicated and inferred reserves of coal and lignites are given in Table 10.3. These cover seams of at least 30 cm thickness down to depths of 1.2 km for coals and 0.5 km for lignite.

Table 10.3. Reserves of coal and lignite

Country	Coal (10^9 tonnes)	Lignite (10^9 tonnes)	Total (10^9 tonnes)	Percentage total
USSR	4121	1406	5527	63.7
USA	1100	406	1506	17.3
China	1011	—	1011	11.6
Europe	143	144	287	3.3
Other	266	85	351	4.0
World	6641	2041	8682	100.0

In the UK it is interesting to note that there is an estimated reserve of 45 000 million tonnes of coal.

What about oil? There are some misconceptions about how oil is found under the Earth. It is not found in great underground lakes; in fact, oil is soaked up by a variety of sedimentary rocks such as sandstone and limestone and its extraction requires the rock to be squeezed like a sponge. Since oil was originally stored in barrels, it has become the practice to use the barrel as the unit of volume when referring to oil; a standard barrel holds 159 litres of oil.

Table 10.4 (references 6 and 12) gives an idea of the worldwide distribution of oil reserves in 1974. It lists **proved reserves** defined as the volume of crude oil which geological and engineering information indicates beyond all reasonable doubt to be recoverable from an oil reservoir under existing operating conditions.

Table 10.4. Distribution of proved oil reserves (1974)

Country	10^6 Barrels	Percentage total
Saudi Arabia	164 500	22.98
Kuwait	72 800	10.17
Iran	66 000	9.22
United States	35 300	4.93
Abu Dhabi	30 000	4.19
Libya	26 600	3.72
Nigeria	20 900	2.92
Neutral Zone*	17 300	2.41
UK	15 700	2.19
Venezuela	15 000	2.10
Indonesia	15 000	2.10
Others	236 597	33.00
World total	715 697	100.00

*This information was published in the *Oil and Gas Journal* in 1974. Just prior to that date the Neutral Zone was an area bordering on Iraq, Saudi Arabia and Kuwait.

These figures emphasise the dominance of the Middle East countries: their total proved reserve was 56.43 per cent of the world's total. For the UK, more recent estimates have varied between 10 and 55×10^9 barrels.

The proved reserves underestimate the true recoverable reserves for they do not take into account oil yet to be discovered, or oil that may be recovered by other methods. Many researchers have made estimates of the world's ultimately recoverable crude oil reserves, and their results have converged around 2×10^{12} barrels. This is now generally accepted as being near to the true figure.

How Long will Resources Last?

Having established reasonable estimates of material reserves, let us now turn our attention to the lifetimes of these materials. It is clear that finite resources must ultimately be depleted, and we can use different methods to estimate their lifetime. First we could assume that the rate of consumption of the resource remains constant. For example, in 1980 the rate of consumption of copper was about 9.2×10^6 tonnes per annum, and the reserve was about 200×10^6 tonnes. If this rate of consumption were to remain constant then the reserve would last $200/9.2 = 21.7$ years. But Fig. 10.3 shows that the rates of consumption do not remain constant. Indeed, they tend to grow exponentially, and over the last 80 years the rate of consumption of copper has been increasing at an average of 4.1 per cent per annum. So if we take the blackest possible picture and assume that consumption will continue to grow at 4.1 per cent per annum, then our total reserve will be gone with a bang in 1994.

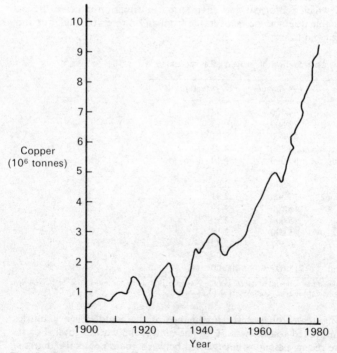

Fig. 10.3. Rate of consumption of copper.

Table 10.5, extracted from *The Limits to Growth*, summarises the results for the resources we have discussed earlier. It is important to note that these are 1970 figures. Here column 3, the **static index**, is the lifetime assuming a constant rate of usage (1970 value). Column 4 gives the average projected growth rate. Column 5, the **exponential index**, is the lifetime if the rate of consumption grows at the value given in column 4. Column 6 gives the lifetime if the reserves were five times greater than that given in column 2. You will note that the reserves in column 2 are less than the more recent ones we referred to earlier (Table 10.2) but column 6 shows how little effect variations in the estimated reserve have on the expected lifetime if the rate of consumption grows exponentially.

Table 10.5. Consumption of selected resources

(1) Resource (1970 value)	(2) Reserves (tons)	(3) Static index	(4) Growth rate	(5) Exp. index	(6) Exp. index with 5 times reserves
Lead	9.1×10^7	26	2.0	21	64
Nickel	6.55×10^7	150	3.4	53	96
Copper	3.08×10^8	36	4.6	21	48
Iron	1×10^{11}	240	2.3	93	173
Aluminium	1.17×10^9	100	6.4	31	55
Coal	5×10^{12}	2300	4.1	111	150
Oil	455×10^9 (bl)	31	3.9	20	50

In practice, however, one would expect the pattern of consumption to lie somewhere between the constant rate of column 3 and the exponential growth rates of columns 5 and 6.

The Pollution Constraint

The Earth cannot go on indefinitely acting as a dustbin. It must have a finite capacity for absorbing pollutants but unfortunately, and this is the dangerous thing, we do not know what the limit is.

The ecosphere, our environment, comprises the Earth's living things and the thin global layer of air, water and soil which is their habitat. Many **ecological cycles** interact to drive this large and intricate system. One, for example, is the water cycle: fish excrete organic waste which is converted to inorganic products by bacteria; in turn these inorganic products promote the growth of algae which are eaten by small aquatic animals who in turn are eaten by fish; and so on. Such cycles purify the environment, the wastes produced in one step becoming the raw materials for the next. They depend on a delicate balance of interaction, and can be upset or destroyed by external agents. For example, if the water cycle is overloaded with organic waste, the bacteria may not have enough oxygen to allow them to do their job of decomposition. Many of our great lakes suffer nowadays from eutrophication, or too much food. First organic waste, in the form of sewage, provides nitrates and phosphates. Secondly, since most lakes are drainage basins, the concentration of nitrates and phosphates can be supplemented by the run-off of dissolved artificial fertilisers from agricultural land.

Scientists calculated that Lake Erie had received 82 500 tonnes of nitrogen in 1968, 45 000 tonnes from sewage and 37 500 tonnes from run-off.

Pollution of lakes and rivers is no new thing. The disgraceful state of the city of Cologne and the River Rhine inspired Samuel Taylor Coleridge (1772–1834) to write these lines:

> In Kohln, a town of monks and bones
> And pavements fang'd with murderous stones
> And rags, and hags, and hideous wenches;
> I counted two and seventy stenches,
> All well defined, and several stinks!
> Ye Nymphs that reign o'er sewers and sinks,
> The River Rhine, it is well known
> Doth wash your city of Cologne;
> But tell me, Nymphs, what power divine
> Shall henceforth wash the River Rhine?

We cannot afford to continue to pollute our planet in this way. A lot of the material we dump will never break down and will not be able to take part in an ecological cycle. Aluminium cans and glass beer bottles are typical examples. Changes in packaging techniques have led to the introduction of the non-returnable bottle and to an enormous amount of plastic, which is not biodegradable.

In addition to all of this rubbish, there are other pollutants that act in more subtle ways. Take carbon dioxide, for example. At present about 95 per cent of our industrial energy comes from the fossil fuels which, when burnt, liberate vast amounts of carbon dioxide. This year more than 2×10^{10} tonnes of this gas will be released into the atmosphere, and over the last century it has been estimated that more than 36×10^{10} tonnes have been generated. Although men and animals contribute by respiration to this total, our machines are about eight times more culpable. What happens to all of this gas? It is known that the ocean absorbs about one half but the rest remains in the atmosphere and can affect the climate. The Earth's cloak of carbon dioxide can act like a greenhouse, allowing the entry of heat in the form of short-wave radiation, but not permitting the exit of the long-wave heat generated by the Earth. The 3.5°C increase in the average temperature of the US between 1920 and 1950 has been attributed to this cause.

Whilst this accumulation of carbon dioxide can lead to increased temperatures, the increasing amounts of dust in the atmosphere can give a completely opposite effect. Average temperatures are now decreasing in many parts of the Earth, and this is explained by the scattering of the incoming solar radiation by dust particles in the atmosphere. Recent measurements have shown large increases in atmospheric turbidity. Examination of dust trapped in frozen snow in the Caucasus has allowed Russian scientists to trace changes in atmospheric dust back as far as 1790, and they found large increases in turbidity after the communists began to industrialise Russia. The world's annual discharge of particulates to the atmosphere has been given as 1600×10^6 tonnes, including 360×10^6 tonnes of naturally-formed sulphates, 200×10^6 tonnes of man-made sulphates and 40×10^6 tonnes of soot.

It has been estimated that a 3 per cent decrease in the transparency of the atmosphere could reduce the Earth's surface temperature by 0.4°C, and although this does not seem much we should bear in mind that the last Ice Age had temperatures only 5° to 8°C below present norms. It would only take a drop of 2°C to cause glaciers to form in the Scottish Highlands. On the other hand, if the **greenhouse effect** were to dominate and temperatures were to rise, then the melting of the polar ice cap could cause the sea level to rise about 70 metres.

Los Angeles is a 'heat island' that has experienced some of the effects of overheating. At present the waste heat or thermal pollution from the high-energy production in the area amounts to 5 per cent of the absorbed solar radiation. It is expected to get worse.

Conclusions

Space has only allowed a cursory examination of the three major threats to mankind: growth of population, depletion of natural resources, and pollution. Is the picture as black as we have painted it? What can we do about it? Whether or not we accept the prophecies of the doom watchers it is important that the technologist examines his actions in the light of these potential threats. He must be aware of the total impact of his actions, for history has shown that what Man aims for and what he gets are often two different things. He builds motorcars to get around more quickly but ends up with traffic jams. He sprays pesticides to control insects but kills birds and fish as well. And this lack of awareness of the total system may drive us, in our quest for growth, to catastrophe. There is a considerable thrust in Denis Gabor's statement that growth addiction is the unwritten and unconfessed religion of our times. Do we need all of our consumer goods? Should quantitative growth give way to qualitative growth? These are questions that each of us has to face up to.

There is, of course, the argument that the prophets of doom always neglect the power of technology, and that in due course a technological solution will be invented for every problem. Taking the three threats in turn, technology might provide the following solutions:

(*a*) **Population:** Methods of birth control are improving and the pill, the intra-uterine device and sterilisation are only a start along this road.

(*b*) **Resources:** Technology will help us to discover new deposits of raw materials, and will give us methods for exploiting low-grade ores. Recycling of materials, and the replacement of scarce materials by common ones will help to maintain our resource bank. Nuclear fission and fusion, backed up by power from wind, water, sun and the tides, will meet our energy needs into the distant future.

(*c*) **Pollution:** Technology has already developed methods of reducing pollution from practically every source to negligible proportions. It only needs the governments of the world to enforce the application of these methods by industry.

Now, you will find that for every argument that pro-technologists present, there will be at least one counter-argument from the anti-technologists, and in order to engage usefully in this debate you will need to be informed.

It is hoped that this book and the reference material will help you in this task.

But it is now time to move on to the second main topic in this chapter: that concerned with technologies for the underdeveloped countries.

10.3   Appropriate Technology

Approximately ninety nations, containing three quarters of the world's population, can be described variously as emerging, **Third World**, developing or underdeveloped. The people of these regions have low incomes and low standards of life; indeed, about 1000 million live in conditions of extreme poverty. The question is often asked: If modern science and technology can put men on the moon, why should it not be able to alleviate the poverty that is suffered by so many of the Earth's inhabitants?

A lot of effort has been directed to solving this problem, particularly over the last twenty to thirty years, and it has become increasingly apparent that the problem faced by the developing countries is the choice of a technology that is appropriate to their needs. The industrially developed nations have an abundance of technological knowledge, and there is a sincere desire to transfer this knowledge to the poorer nations, but unfortunately much of our advanced technology would be inappropriate to the needs of these nations. In order to clarify this, let us try to identify the main features of Western technology.

Basically the advanced countries are centred upon, and are moving towards, a capital-intensive technology that uses lots of expensive machinery and employs only a little labour. These industries rely on an intricate infrastructure, including specialised education to high standards, a disciplined labour force, a far-reaching transport system, specialised fuels and a complicated system of social services. In addition, many of these industries rely on mass markets and tend to be based in or near large cities.

When we think for a moment about the needs and the infrastructure of the developing countries, it becomes quickly obvious that the direct transfer of Western technology into a developing country would be neither effective nor desirable. The developing countries are poor, and they suffer from high birth rates, high unemployment, bad transport systems and communications, lack of adequate water and electrical supplies, and inadequate health and education services. Their economics are largely agrarian. About 80 per cent of their population lives in well over a million rural villages, and most of them practise traditional methods of agriculture on small land holdings.

So we can see many reasons why advanced technology is unlikely to be appropriate for the developing countries. Let us list a few:

1. Western technology is expensive.

2. Capital-intensive industries would not provide the millions of jobs required to reduce unemployment to reasonable proportions.

3. The highly skilled worker and managers could not be provided by most of these countries, so staff would have to be imported.

4. The products of Western technology would be in most cases only

saleable abroad. Even if they were suitable for the home market, their availability would inevitably lead to the unemployment of local people who make the products by traditional methods.

There are many examples of the failure of attempts to implant advanced technology in the developing countries. In one case a North African country decided to establish a highly automated plant, including plastic moulding injection machines, to produce plastic sandals. The traditional sandal had been made of leather, but it was felt that the new product would be cheaper and longer wearing. However, the outcome was that 5000 shoemakers were made redundant and this in turn reduced the income of a whole series of other people such as leather makers, glue manufacturers and toolmakers. In addition, the plastic and the spares and maintenance had to be imported. This technology was clearly not appropriate to that country at that time.

These are some of the reasons that led Dr E. F. Schumacher to the concept of an 'intermediate' technology. In 1963, he visited India where he was extremely impressed by the Gandhian ideas of rural technology and industrialisation. He first used the name **intermediate technology** in a paper presented to the Indian Government in 1964, and an expanded account is given in the book *Small is Beautiful*. Schumacher's definition of intermediate technology was 'a technology of production by the masses, making use of the best of modern knowledge and experience, conducive to decentralisation, compatible with the law of ecology, gentle in its use of scarce resources, and designed to serve the human person instead of making him a servant of machines'. It was intermediate in that it stood somewhere between the low and the high technologies (Fig. 10.4). For example, the technology of the ox plough would be considered to be intermediate to the hand-operated hoe and the modern diesel-driven tractor. Thus the ox plough would be considered an intermediate technology in parts of Africa. But the word 'intermediate' is a relative one, and what is an intermediate technology for one country could be a low technology for another. For example, the ox plough has been used for thousands of years in the Middle East and Asia, and in these countries small two-wheel tractors, such as those developed by the International Rice Research Institute, would be an intermediate technology.

The term **'appropriate' technology** is nowadays thought by many people to reflect more accurately their concern for the well-being of the underdeveloped countries. The name emphasises the social and cultural dimensions as well as the technological, and so it has tended to replace the older intermediate technology concept. We have already discussed some aspects of appropriate technology, and what makes a technology appropriate to a particular set of circumstances. A list of criteria of appropriateness has been developed by the Brau Research Institute in Canada, and this is given below:

(*a*) Appropriate technology should be compatible with local cultural and economic conditions, i.e. the human, material and cultural resources of the community.

(*b*) The tools and processes should be under the maintenance and operational control of the population.

Simple

Intermediate

Advanced

Fig. 10.4. Levels of technology.

(*c*) Appropriate technology, wherever possible, should use locally available resources.

(*d*) If imported resources and technology are used, some control must be made available to the community.

(*e*) Appropriate technology should, wherever possible, use local energy resources.

(*f*) It should be ecologically and environmentally sound.

(*g*) It should minimise cultural disruption.

(*h*) It should be flexible in order that a community should not lock itself into systems which later prove inefficient and unsuitable.

(*i*) Research and policy action should be integrated and locally operated wherever possible, in order to ensure the relevance of the research to the welfare of the local population, the maximisation of local creativity, the participation of local inhabitants in technological developments and the synchronisation of research with field activities.

These are admirable aims, and indeed one would wish to see similar aims applied to our Western technologies. They really get to the heart of matters. Is there a real need? How can we meet it and at the same time provide the maximum benefit to the community? You will recall our discussion of criteria in Chapter 7 on the methodology of technology. The above list should help us to identify constraints and criteria when comparing various solutions to a problem, not only in the underdeveloped countries. Indeed, it could be argued that the appropriate technology movement is forcing us to reassess our own technologies.

All of the above has been concerned with the philosophy behind appropriate technology. Let us now spend some time on specific examples related to water engineering.

Appropriate Technology in Water Engineering

As individuals we need about 2 to 3 litres of water per day to sustain our bodily processes, and of this we would normally get about 60 per cent from our food and 40 per cent from drink. This is the absolute minimum requirement, but in the UK, for example, each individual uses about 200 litres per day for domestic use alone. This covers washing, cooking and so on; indeed the flush w.c. can account for about one third of the total. The total figure grows to about 300 litres per day per individual if we take into account the needs of industry. So the availability of vast amounts of pure water for human, animal, agricultural and industrial purposes is taken for granted in the developed countries.

The problems of water supply and control are very serious in the developing countries, most of which lie in the tropics or subtropics where climates consist of wet and dry seasons of unpredictable severity. These can lead to droughts and floods. In the dry seasons surface resources can dry up, and there is a need for a system of water conservation and storage to ensure a reliable supply during the dry season. The concept of the rainwater catchment tank is not new, but nowadays with the advent of polythene and other synthetic materials it is possible to build small low-cost tanks for use in rural areas. The Intermediate Technology Development Group have developed designs for these tanks for Sudan, Botswana, Swaziland and Jamaica, The basic material is thin polythene sheet and tubes of about 8 cm in diameter (Fig. 10.5). A pit is dug and the walls and floor are

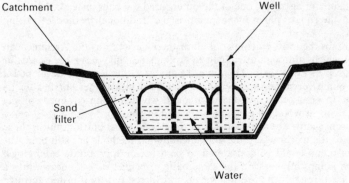

Fig. 10.5. Water reservoir: catchment tank.

first lined with alternate layers of mud and polythene sheet and finally with cement filled polythene tubes. This method of construction allows the builder to use only about a quarter of the amount of cement required for normal concrete blocks. A sand filter is used to purify the water, and this requires the construction of several 'beehive' domes using the cement-filled tubes. These domes collect the purified water. The completed well has to be covered to avoid contamination and evaporation. A tank of 40 000 litres capacity can be constructed from polythene costing about £20, and requires about 120 man-hours of unskilled labour.

The problem of disease is very serious in the developing countries and water and health are closely related. The World Health Organisation have estimates that 80 per cent of all diseases and illnesses are carried by contaminated water. Infections such as typhoid and cholera can spread through water supplies. Others such as guinea worm are transmitted by aquatic animals such as snails. And others such as malaria and sleeping sickness are spread by insects that depend on water. So it is extremely important that purified water sources, such as the catchment tank, are introduced as widely as possible.

Successful water engineering also requires efficient and reliable pumping systems. Some of the traditional water-lifting systems are shown in Fig. 10.6, and you will notice that they are all types of bucket which are filled, lifted and emptied. The scoop and the shaduf both employed the principle of the lever, but the shaduf was particularly effective. It was used by the Egyptians to lift water from the Nile for irrigation purposes and it has been estimated that in one day a man would have been capable of raising about 2500 litres to a height of 2 metres. The first serious rival to the shaduf was the Persian wheel or saqiya, that was reputed to have raised water from a well beneath the Hanging Gardens of Babylon. It was used in Egypt around 200 BC. The dragon's spine was basically a Persian wheel inclined to the vertical, but instead of using buckets it used flat paddles to push water up a trough. It is still used in China.

The origins of the chain pump are somewhat obscure but it must go back a long way in history, for Agricola described no less than six different varieties in his book *De Re Metallica*, published in 1556. Fig. 10.6 shows the chain and disc type; one of the others used rag balls instead of discs. In China the square pallet chain pump is the traditional method of raising water.

Perhaps the most ingenious of all of these devices was the Archimedean screw which, although attributed to Archimedes, may have been used for irrigation in Egypt before his time. In any case, the Romans developed it as a pump for mines and renamed it the Cochlea (water snail) after its shape. In practice it consisted of a round wooden beam wrapped spirally with strips of wood on edge, then encased in boards.

Piston pumps are now popular in the developed countries, but in those parts of the world where the above traditional methods are still used, the piston pump could be considered to be a useful appropriate technology. Piston pumps all work on the same basic principle: water is drawn into a cylinder through a non-return valve, and forced out by a piston through

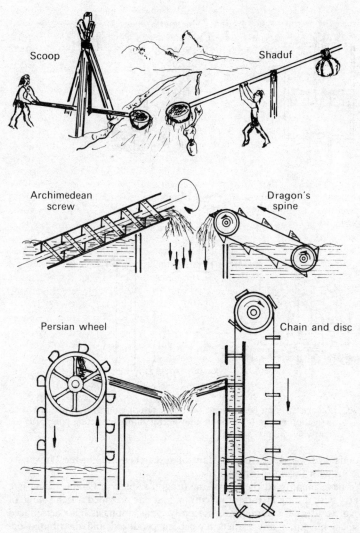

Fig. 10.6. Water lifting.

another non-return valve. Fig. 10.7(*a*) will help to clarify the operation of the simplest village pump. The piston descends and water is forced to pass upwards through the non-return valve in the piston. When the piston is forced upwards it lifts the water to the outlet and, at the same time, it draws water through the bottom non-return valve.

An interesting variant on this theme is shown in Fig. 10.7(*b*) which shows how an old tyre can be used as an irrigation pump. The Third World

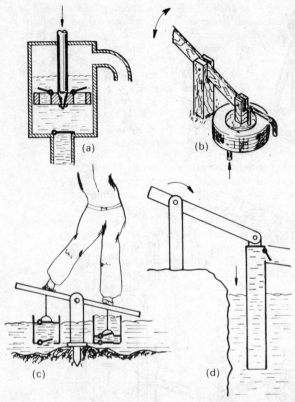

Fig. 10.7. A variety of low cost pumps: (*a*) village piston pump; (*b*) a tyre used as an irrigation pump; (*c*) a simple foot-operated, double-acting pump; (*d*) the flap-valve pump.

abounds with old tyres, and a team of researchers at Purdue University came up with this idea for their use.

We have mentioned the use of the dragon's spine in China. China is a country with a long history of invention, and has provided a rich source of innovative ideas for agriculture. Recently great emphasis has been placed on self-help, and many leaflets have been produced and distributed describing simple tools and machines. Several of these relate to irrigation, and one, a simple double-acting lift pump, is becoming popular (Fig. 10.7(*c*). It has square section cylinders, can be constructed from wood by the village carpenter, and is driven by a standing operator shifting his weight from one foot to the other. A similar pump, but using canvas bellows instead of pistons, has been developed by the International Rice Research Institute in the Philippines. The mass of the unit is only 20 kg, and it can lift about 250 litres a minute to a height of 1 or 2 metres. Its main use is to raise water from shallow ditches to the fields.

The last device we have space to describe is the flap-valve pump which, as you will see from Fig. 10.7(*d*), has only one non-return valve. It is extremely simple; none of the dimensions is critical, and it can be constructed from locally available materials. Its action relies on the inertia of the vertical column of water. When the pipe descends the water remains relatively stationary and leaves through the spout. When the pipe is raised the whole column of water is raised with it. An even simpler version does away with the flap valve and replaces it with an operator's hand! The flap-valve pump has demonstrated a capability of lifting about 120 litres per minute through 4 metres using a pipe with a diameter of 8 cm.

There are many ways of improving the performance of these simple pumps, and there are many other types of more advanced pumps that could be adapted for use in the Third World. The Afghan joggle pump is an excellent example of the former: it is an improvement on the flap-valve pump, using a trapped air volume to store energy. Examples of more advanced pumps are the hydraulic ram which uses the 'water hammer' effect for pumping, and the Humphrey pump which is essentially an internal combustion engine that uses the column of water as its piston.

It is interesting to note that several appropriate technology projects have involved the introduction of wind power to traditional man-operated pumping systems. One scheme, recently used for raising water in a village community in Zambia, uses a Savonius rotor made from an oil drum cut in half vertically. The two halves are mounted on a vertical axis (Fig. 10.8) and the rotor rotates at low speed with a good starting torque. A great advantage is that it does not have to be orientated with respect to the wind.

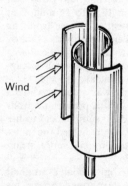

Wind

Fig. 10.8. The Savonius rotor—a simple windmill with a vertical axis.

In concluding this section on appropriate technology, I would wish to re-emphasise that there is more than technology involved in the concept. The successful introduction of any new technology requires that it be acceptable to the community, that they understand how it works, why it is an improvement and why they should bother to be responsible for it. The following extract from an article on 'Water for India', in the *New Internationalist* encapsulates this view:

'A truck appears, a 150 foot well is sunk, a hand pump is installed, cameras click, speeches are made and the truck is cheered on its way to the next village. The villager often does not know how the rig and the pump work, where they have come from, why they have come to this particular village, what advantages they have to offer. He has not contributed anything in the planning, money, labour or time. He is uninvolved. And then it is taken for granted that he will feel responsible for it.'

Let us now move on to the concluding section of this chapter—that concerned with the impact of automation on society.

10.4   Men and Machines

There has always been a love–hate relationship between Man and his machines. He loves them because they make life less arduous. He hates them because of the social upheavals they have created, and he is frightened of the way in which, at the touch of a button, they could wipe mankind off the face of the Earth. And there is a growing fear that machines with artificial intelligence will soon take over the world.

The Industrial Revolution

Let us first look at the social upheaval that a new technology can generate, and where better to start than with the Industrial Revolution, covering a period roughly from the middle of the eighteenth to the middle of the nineteenth centuries. During that period there was an enormous expansion in Britain's population; it grew from 6.5 million in 1750 to 14 million in 1831. At the same time there was a great movement of people from the countryside to the new industrial cities; in 1790 there were twice as many people in the countryside as in the towns, but in 1840 this was almost completely reversed. And the result of all of this new industry was a staggering increase in the number of machine-made products.

It has been argued that the Industrial Revolution was a complex inter-relationship among developments in coalmining, steam power and iron-working. It was necessary for the collieries to produce enough coal to fire the steam engines and to melt the iron. The steam engine helped the mines to produce more coal and the iron was used to produce more steam engines—and so on.

Let us look more closely at this important threesome—coal, steam and iron. For centuries, coalmining had been left to the farmer to turn his hand to when sowing and harvesting had been completed. However, as industrial demand grew mines had to be sunk deeper and deeper, and the task soon became more than a part-time job. Up until the end of the nineteenth century the actual business of cutting the coal from the pit face was a matter of human muscle power, assisted occasionally by blasting powder. Transportation of the coal from the pit face to the pit head also required brute force, and this was often carried out by women and children. This was a terrible task; women and children had to crawl on their hands and knees in cramped, dark tunnels, dragging heavy sledgefuls of coal behind them.

Steam was another of the threesome, and we have already seen in Chapter 1 how Savery's steam engine 'miner's friend' was used to pump water from mines in 1698, and led in turn to Newcomen's engine in 1712 and then to Watt's engine in 1776. James Watt, in collaboration with Matthew Boulton, the financier and organiser, produced an engine that was an immediate success, with miners, iron-masters and manufacturers queuing to place orders.

We also touched on the development of iron-working in Chapter 1. The annual production of iron in Britain at the beginning of the eighteenth century has been estimated at 17 000 tonnes, but by 1790 this was to grow to 68 000 tonnes and in just another 10 years it had shot up to 125 000 tonnes. This phenomenal growth was to continue as industrial output grew, and by 1825 it had exceeded half a million tonnes.

For many centuries charcoal had been used as the fuel for smelting the iron, and as a result much of the early iron industry was concentrated in the Sussex Weald where both wood and water were plentiful. However, as wood ran out and coal became more plentiful, it was not surprising that the iron producers began to experiment with coal instead of charcoal. In 1709 Abraham Darby turned coal into coke that allowed him to produce an iron suitable for casting saucepans, kettles, railway lines and bridges. But it was not until 1784 that a superior type of iron, less brittle than cast iron, was produced. This malleable iron could be milled, wrought and turned, and was produced by a technique called 'puddling' which required the molten iron to be continuously stirred to burn off the sulphur contained in the coal. The removal of the sulphur was one of the factors essential to the iron's excellence, but the physical task required of the men, called puddlers, was gargantuan. They had to stir the molten metal continuously with long iron bars.

Thus we can see an abundant evidence of a lack of concern for human labour in the developing technologies of coal, iron and steam. But it was not only the excesses of physical labour that made the Industrial Revolution notorious for its insensitivity to human needs. The shift in population from country to town, and the shift from cottage-based crafts to factory-based production lines, caused a tremendous social upheaval. This is clearly illustrated by the changes that took place in the textile industry at that time.

Changes in the Textile Industry

The textile industry was originally a cottage-based industry with the family as the working unit. The cloth merchants or clothiers delivered wool to the cottages, and when it had been turned into cloth they collected it again for distribution to the markets of the world. In the cottages every member of the family had a part to play in the manufacture of the cloth. The wife and children usually looked after the carding or the disentanglement of the wool into separate lengths suitable for spinning; it was a sort of combing operation. The wife would also do the spinning on the old spinning wheel so popular in fairy tales. Finally, the husband wove the continuous thread or yarn into cloth on his loom.

Spinning was a slow process compared to weaving, and this was made even worse in 1733 when John Kay invented the flying shuttle that allowed a seated weaver to pass the shuttle in both directions across a piece of cloth far wider than the span of his arms. Thus the gap between supply and demand of yarn became wider. It was not long, however, before inventors had devised ingenious ways of bridging the gap.

The first was James Hargreaves who lodged a patent for his spinning jenny in 1770. The device was named after his wife, and tradition has it that the idea was inspired by observation of the way a spinning wheel ran on after it had been accidentally knocked over. Basically the jenny extended the simple spinning wheel by using one treadle to operate eight spindles and so increased the output of one woman by a factor of eight.

Around the same time Richard Arkwright, a barber by trade, manufactured a spinning machine that was capable of producing a strong cotton yarn ideal for weaving the cotton that was to be the material of the future. Then along came Samuel Crompton who, between 1774 and 1779, developed a machine that used the best features of both Hargreaves' and Arkwright's machines. Because of its hybrid nature it came to be known as Crompton's mule.

Now these machines certainly increased productivity in the textile industry, but their introduction heralded the beginning of a century of misery and deprivation for the workers; a century that was to see the demise of the cottage industry and the rapid growth of the factory system.

At first the handloom weavers did well. They were in great demand for the production problem had now been reversed and yarn was being produced more quickly than it could be woven. Although the spinning jenny could be installed and operated in the cottage, Arkwright's machine and Crompton's mule needed a lot of power and had to be operated under factory conditions. From 1790 onwards the use of steam power spread simultaneously with the expansion of the cotton industry, and more and more yarn was being produced. But this Golden Age of the handloom weaver was to be shortlived.

The invention of the power loom sprang from the needs of the cotton industry. Edmund Cartwright foresaw the tremendous growth in spinning mills and saw the need for power looms that could cope with this rapidly increasing output of yarn. Mechanisation of the loom was not an easy task, but Cartwright had achieved a workable solution around about 1790, although it had many imperfections and required further refinements by other inventors before being accepted in large numbers in the 1820s. By 1850 the cotton industry had shown an enormous swing to power looms, nearly a quarter of a million being in operation in factories, compared to about one fifth that number of handlooms. This was to cause a great upheaval in the lives of the handloom weavers.

The handloom weaver's Golden Age was at an end, and whereas he could command an income of around £5 a week in 1790, by 1830 he was lucky to earn ten shillings (£0.50) for a week of 18-hour days. He was in a dreadful dilemma. If he worked at home he had to work hard for the low wage in order to compete with the mass production of the factories. But if he thought of throwing in his home job and seeking employment in the factories, he was in for another disappointment. The newly mechanised looms did not need the physical strength of a man; rather they required the dexterity and nimbleness of women and children for jobs such as repairing broken threads. This was to turn family traditions on their heads: the women and children would have to be the breadwinners.

In addition to this aspect of social upheaval we should also note the dreadful conditions that these women and children had to face in the mills. Work would start around 5 o'clock in the morning and would continue, with perhaps one or two short breaks for meals, until around 8 o'clock in the evening. Working conditions could be appalling; factories were noisy, dirty and boring; and the workers' health often suffered.

Reactions against the Machines

We have seen that the Industrial Revolution caused a whole series of related disturbances: the reduction in craft skills; the move from cottage to factory; the move from the country to the town; the employment of women and children; the upset of family life. It was not surprising therefore that people came to fear the machine, and in due course the workers and the intellectuals combined to raise their voices against its evils. The so-called Luddites (1811–17) were a group of textile workers who destroyed tens of thousands of pounds worth of machinery in their anger and frustration at their social conditions. And around 1830, agricultural workers, led by the notorious Captain Swing, took part in many violent demonstrations against their state of poverty. One of their attacks was concentrated upon the threshing machines that were removing their livelihood.

The intellectuals made their contribution too. In 1829 Thomas Carlyle predicted unemployment when he wrote: 'Our old modes of exertion are all discredited and thrown aside. On every hand the living artisan is driven from his workshop to make room for a speedier inanimate one.' In the late nineteenth century Samuel Butler predicted that the day would come when 'Man shall become to the machine what the horse and dog are to us'.

This reaction from society was to lead to wide social reforms and in due course to the foundation of the trade unions. But it is interesting that the same fears—those of unemployment and the rule of the machine—are still with us today. So we shall conclude this chapter by having a brief look at the microprocessor and its impact on employment.

Employment and the Microprocessor

We are now entering yet another industrial revolution. In the first one, machines were developed to replace human muscle power, but in this new revolution machines are rapidly augmenting and replacing human brain power. One of the brain's main functions is to receive, interpret and dispatch information, and this is something that computers are very good at. We are moving towards an information-based society in which the ability to collect, store, organise, analyse and act upon information is going to be a central feature. Information management will be to the twenty-first century what steam, coal and iron were to the eighteenth and nineteenth centuries.

And why is information so important? There is some truth in the old adage that 'ignorance is bliss'. It depends upon what you mean by bliss.

But if we wish to respond to Man's relentless desire to develop, to improve, to inquire, then accurate information has to be available on which to base the sort of decisions that we discussed in Chapter 7. For example, an industrialist needs up-to-date and accurate information on world markets before deciding on whether to manufacture a product; governments need accurate information concerning employment distribution when drawing up economic policies; power engineers need accurate information about power demands when they are trying to optimise the efficiency of electrical distribution networks; and we ourselves would like to have accurate information about the prices of consumer products, like briefcases, so that we can get the best buy.

So information, like energy, is an essential resource for the effective operation of a country's economy. It will become the key resource demoting, in time, the more traditional factors such as capital, labour and raw materials. The Japanese foresaw this early in the 1970s when they first talked about the Information Society. Nowadays the development of information technology and the new science of informatics emphasises the growing importance of this topic.

The microprocessor will take the lead in the Information Society. It has reduced the cost of computing by an enormous factor, so that the range of use of computers is now rapidly expanding. To date computers have been used mainly for handling numerical information, but most of the information we use, e.g. this book, is not numerical but textual—words not numbers. It is not surprising therefore that a lot of effort is now being devoted to the development of computing techniques to process textual information.

Today most of us handle information in the same way. We use paper as the storage medium, post and telephone as transmission systems, pens and typewriters as input–output equipment, and the human brain as the most commonly used processing device. This is all going to change as microelectronics develop to give us a closed system for handling information in the form of electronic signals. It is not surprising therefore that the biggest upsets in employment are expected to occur in those vocations that involve a large degree of information handling.

Over recent years the UK has seen a growing shift of employment from manufacturing industries into the service industries (see Chapter 9). These people—secretaries, sales clerks, stock clerks, accountants, cashiers, administrators, and so on—are basically information handlers. Their days are taken up mainly with paper work, telephone calls and meetings. Recent research by the Science Policy Research Unit at Sussex University concludes that nine million people, or 40 per cent of the working population of the UK, can be categorised as information handlers. So it is clear that we are already rapidly moving towards the Information Society.

It is interesting to note that the job of the 'white collar' worker, or information handler, will be in greater jeopardy than the 'blue collar' worker who is involved directly in the manufacturing process. A lot of money has already been invested in increasing the efficiency of the manufacturing process so the room for improvement in productivity is reduced. Thus, for example, it has been argued that if British industry were to spend about 1 per cent of its total turnover on robots, then about 10 per cent of the five million direct-production workers could be out of a job by the end of the 1980s. This sounds a lot but it is much less than many more hysterical predictions.

Compare these figures with the job losses that may occur in the information handling sector, where there are about nine million employees in the UK. There has been little investment in this area, and the typical information handler has the minimum equipment of a desk, a telephone, a filing cabinet and maybe a typewriter or a copying machine. Increased investment in microelectronic equipment producing, for example, a conservative 20 per cent increase in productivity, would ensure that the same work could be done by nearly two million fewer employees. This is more serious than the situation in the manufacturing industry. But there is always hope, for there are several examples where technological advance has created new jobs to replace those destroyed. Look, for example, at the way in which banking and insurance expanded rapidly when computers were introduced, and how at the same time the number of wage clerks reduced drastically.

Can we predict how the microprocessor is going to revolutionise the work of the information handler? Let us look briefly at the office of the future.

The Office of the Future

An office is basically concerned with the transmission and storage of information. Transmission is usually by telephone or letter, and the filing system is the most common means of storage. The information originating at the manager, for example, could have to traverse the following steps. First it is transmitted to the secretary in handwritten form or by dictation; then it is typed and a copy filed; it is placed in the mail tray; it is collected; it is delivered; it is opened by another secretary; it is read by another manager and then filed. Microelectronics will affect all of these functions.

New methods for carrying out the first step, the transmission of the information to the secretary, are under development. Microcomputers that can interpret handwriting are available, and in the longer term it is highly likely that it will be possible to communicate with computers by voice.

The secretary's basic tool, the typewriter, is being rapidly displaced by the word processor and its advanced cousin the electronic typewriter. The word processor is basically a small computer for processing words rather than numbers. It has a keyboard that the typist uses in much the same way as the conventional typewriter, but in addition it has a visual display like a TV screen as well as a printer for producing the conventional 'hard' copy. The computer's memory allows the typist to correct and edit a document without having to retype the unchanged parts of the text. Once it is completed to everyone's satisfaction it can be reproduced ad infinitum by the printer. In addition, the memory allows many other useful functions such as the addition of names and addresses to circular letters. It is clear that the increasing use of word processors will have a serious effect on the future of the copy typist.

The word processor converts the information to digital form, ideal for transmission as 'electronic mail'. It is not difficult to imagine a large organisation interconnecting the word processors in each of its offices, and perhaps augmenting this with a further computer to provide an integrated communications network or information ring-main. Thus a letter entered on one word processor would be immediately available at the word processor of the addressee. Such systems are already in action and are clearly faster and superior to the conventional 'foot mail'. It is possible to extend such a system beyond the precincts of a particular organisation so that offices around the country, or indeed the world, are connected in this way.

One can also see that the conversion of textual information to digital form offers tremendous scope for improvements in storage. Conventionally this has relied on filing cabinets and paper—vast quantities of it. As our storehouse of knowledge increases the sheer physical bulk of the paperwork grows exponentially; for example, it is estimated that science alone generates about six million new facts every year, each requiring books and learned papers for its dissemination. Storage of this information on microchips and on magnetic tape makes it not only more compact but also more readily available. So the typist's letter, produced on the word processor, can be held in the short term in the word processor's RAM, and for longer terms on magnetic tape.

These few examples illustrate how microelectronics can improve the productivity of the office. They also indicate the sort of changes we can expect to see. Jobs such as the copy typist will be in jeopardy, and others, like secretary and manager, are bound to change in emphasis and content. Many of the elements of the present paper-based information systems will disappear in the near future—the mechanical typewriter, the filing system, the photocopying machine, the postal service.

Conclusions

So, in concluding this section we should emphasise that the fears of today, in relation to the development of machines, are hardly different from the fears that arose during the first industrial revolution. Man looks for security and respect, and his ideal employment has provided him with both. As machines develop and productivity increases we will reach a stage where, although employment is drastically curtailed, sufficient wealth will be available to ensure the security of all. The problem will lie in the distribution of the wealth, but that is a problem for the politicians. Another problem will be the proper use of leisure time, but that one is a problem for the sociologists!

10.5   Exercises

1. The complexity and the rapid growth of technology place it beyond the understanding of the average citizen. What, if anything, should be done about this?

2. (*a*) Choose five products or systems that you consider represent 'good' technology. (*b*) Choose another five for 'bad' technology. Why do you describe them as good or bad? Would your friend agree?

3. Write 500 words on who should be blamed when technology turns out to be bad.

4. (*a*) What is the world's reserve of oil? (*b*) List the top four producing countries. (*c*) List the top four consumer countries. (*d*) How long will reserves last? (*e*) What should be done to make these last longer? Discuss the ramifications of your proposals.

5. (*a*) What was the Club of Rome? (*b*) Their study recommended a move towards a zero growth economy. What does this mean? How would it be implemented?

6. (*a*) Discuss the major factors contributing to the growth of the world's population. (*b*) What did Malthus have to say on this topic? (*c*) Describe four ways of reducing population growth, commenting on practicality and side effects.

7. Examine the use of water in your home. Determine (*a*) the various major uses and their consumption and (*b*) the daily use per person. Suggest steps to reduce the water consumption by a half.

8. The depletion of energy resources has led to proposals for homes that are self-sufficient in energy. Describe how you would imagine such a house to operate.

9. (*a*) What is the Third World? (*b*) List 20 countries that belong to the Third World. (*c*) What do you consider to be the five major needs of these countries?

10. (*a*) What is Alternative Technology? (*b*) Write 50 words each on the application of Alternative Technology to agriculture, housing and transport.

11. Describe five forms of energy generation that would be of value to a village in an underdeveloped country. How would you persuade the villagers to accept these technologies?

12. Write 500 words describing the good and bad features of the Industrial Revolution.

13. Describe the changes that have occurred in the pattern of work on farms as a result of mechanisation and increased dependence on fossil fuels.

14. Give examples of Man's reactions against the machine, from the Industrial Revolution to the present day.

15. Some people claim that computers destroy jobs—others that they create jobs. Give examples supporting both sides.

16. There will be more leisure time in the future. Suggest ways of putting it to good use.

10.6 Further Reading

1. Braun, E., and Collingridge, D., *Technology and Survival*, Butterworths, 1977.
2. Cole, H. S. D., *et al.*, *Thinking about the Future*, Chatto and Windus, 1974.
3. Congdon, R. J., *Introduction to Appropriate Technology*, Rodale Press, Emmaus, Pennsylvania, 1977.
4. Dunn, P. D., *Appropriate Technology*, Macmillan Press, 1978.
5. Ehrlich, P. R., and Harriman, R. L., *How to be a Survivor: A Plan to save Spaceship Earth*, Pan/Ballantine, 1971.
6. Foley, G., *The Energy Question*, Pelican, 1976.
7. Hales, M., *Science or Society?*, Pan Books with Channel Four Television Company, 1982.
8. *Living with Technology*, The Open University, T101 Block 4, 1980.
9. *Maintaining the Environment*, The Open University, T100, 26 and 27 1972.
10. Mason, B., *Principles of Geochemistry*, Wiley, 1966.
11. Meadows, D., *et al.*, *The Limits to Growth*, Earth Island Ltd and Pan Books, 1974.
12. *Oil and Gas Journal*, December 30, 1974.
13. Papanek, P., *Design for the Real World*, Paladin, 1974.
14. *Resources and Man*, W. H. Freeman & Co., 1969.
15. Saury, A., *Zero Growth*, Basil Blackwell, 1975.
16. Taylor, G. R., *The Doomsday Book*, Panther Books, 1972.
17. *Technology for Development*, Centre for World Development Education, 1977.
18. United Nations Statistical Year Book, 1972.

Appendix 1
The SI system

Multiplier	Prefix	Symbol
10^{12}	tera	T
10^{9}	giga	G
10^{6}	mega	M
10^{3}	kilo	k
10^{2}	hecto	h
10	deca	da
10^{-1}	deci	d
10^{-2}	centi	c
10^{-3}	milli	m
10^{-6}	micro	μ
10^{-9}	nano	n
10^{-12}	pico	p
10^{-15}	femto	f
10^{-18}	atto	a

For example

$$1\ \text{MW} = 1\ \text{megawatt} = 1000\ \text{kW}$$
$$1\ \text{cm} = 1\ \text{centimetre} = 10\ \text{million nm}$$
$$1\ \text{kg} = 1\ \text{kilogram} = 1000\ \text{million } \mu\text{g}$$

And, on a humorous note

$$1\ \text{million phones} = \text{a megaphone}$$
$$\text{One millionth of a phone} = \text{a microphone}$$

And even worse

$$10\ \text{cards} = \text{a deca cards}$$
$$\text{and } 10^{12}\ \text{bulls is terabull!}$$

For even more outrageous examples see Philip Simpson's article 'Standards for inconsequential trivia' in the book *A Random Walk in Science* (Institute of Physics).

Appendix 2
Answers to Selected Exercises

Chapter 2

1(b) 39.24 kJ, 1(c) 261.6 W.   2(c) 20 mm, 0.2, 5.0, 80 N.   4(b) 196.3 kW.   7(a) (i) 59.71 Nm, (ii) 200 rev/min, 238.84 Nm. 7(b) 800 rev/min, 47.77 Nm   9(b) 174.4 N, 9(c) 19.62 kJ, 9(d) 98.1s., 9(e) 1.44 l/s.

Chapter 3

12(b) (i) 7 bits, (ii) a further 25 bits, (iii) about another 16 bits, 12(c) 75 bits.

Chapter 4

7(b) 1.62.   8(a) $c = 8u + 10 r$, 8(b) $A = 50$, $H = 0.09$.   10(d) 0.429, 0.5, 0.524.

Chapter 5

3(b) 1000111, 3(c) 1100011, 3(d) 0010001, 3(e) –001111.   6. A decimal to binary encoder.   9 and 10 The machine code and assembly code are as follows:

01 000	CLA
06 002	INA 002
03 109	INP 109
03 110	INP 110
05 111	TR 111
01 000	CLA
08 109	ADD 109
08 110	ADD 110
10 111	DIV 111
05 112	TR 112
04 112	PT 112
02 000	STOP

11(c)

```
10   INPUT X, Y
20   PRINT X * Y/2
30   END
```

12(b) The BASIC program is as follows:

```
10   INPUT B,C
20   A = B*B − 4*C
30   IF A >= O THEN 60
40   PRINT "NO REAL ROOTS"
50   END
```

```
60   Y = 0.5    *(SQR(A) − B)
70   Z = −0.5 *(SQR(A) + B)
80   PRINT Y,Z
90   END
```

Chapter 6

4(*a*) 0.604 m/s², **4**(*b*) 49.67 s, **4**(*c*) 745 m.   **6**(*b*) 34.3 years, **6**(*c*) 144 years.   **7**(*c*)(i) 160,000, **7**(*c*) (ii) 193 152 (new population is 742 892).   **9**(*a*) 10 000 m², **9**(*b*) 142.8 m³ (volume is maximum when radius $r = \sqrt{A/3\pi}$, where *A* is total area of plastic).   **10.** Optimum mix is 480 record players and 540 televisions; Dept. A is not fully utilised. Equations are:

$$8r + 10t < 12\ 000$$
$$4r + 12t < 8\ 400$$
$$12r + 16t < 14\ 400$$
$$\text{profit} = p = 20r + 40t$$

11. Design, buy part A, assemble, test, paperwork = 30 weeks.   **12**(*a*) AJGF = 350 km, (*b*) AJDEF = 360 km.

Chapter 7

7(*a*) M1 = 225, M2 = 257, M3 = 275, **7**(*b*) M1 = 1.125, M2 = 1.512, M3 = 1.486, **7**(*c*) M2. **8**(*a*) *A* = 2 yrs, *B* = 2.5 yrs, *C* = 3.2 yrs, **8**(*b*) £1300, £9055, £9800, **8**(*c*) C.

Index